Advances in Road Pavements

Advances in Road Pavements

Guest Editors

Emanuele Toraldo
Misagh Ketabdari

MDPI

Basel • Beijing • Wuhan • Barcelona • Belgrade • Novi Sad • Cluj • Manchester

Guest Editors
Emanuele Toraldo
Department of Civil and Environmental Engineering
Politecnico di Milano
Milan
Italy

Misagh Ketabdari
Department of Civil and Environmental Engineering
Politecnico di Milano
Milan
Italy

Editorial Office
MDPI AG
Grosspeteranlage 5
4052 Basel, Switzerland

This is a reprint of the Special Issue, published open access by the journal *Buildings* (ISSN 2075-5309), freely accessible at: https://www.mdpi.com/journal/buildings/special_issues/Q7D4G415Z9.

For citation purposes, cite each article independently as indicated on the article page online and as indicated below:

Lastname, A.A.; Lastname, B.B. Article Title. *Journal Name* **Year**, *Volume Number*, Page Range.

ISBN 978-3-7258-6698-4 (Hbk)
ISBN 978-3-7258-6699-1 (PDF)
https://doi.org/10.3390/books978-3-7258-6699-1

Contents

About the Editors

Emanuele Toraldo

Emanuele Toraldo is an Associate Professor in the field of Roads, Railways and Airports. Since 2001, he has been actively engaged in research and teaching related to road and airport materials, pavement assessment and road design, construction technologies, and maintenance strategies. His research activities focus on pavement performance, surface characteristics, durability, and innovative materials for transportation infrastructures. He is a member of several national and international scientific boards and technical committees and has contributed to research projects at both national and European levels. He is co-inventor of several Italian and European patents and the author of two Italian books on road infrastructure engineering. In addition, he has published more than 60 scientific papers in peer-reviewed journals and international conference proceedings.

Misagh Ketabdari

Misagh Ketabdari is an Assistant Professor in Transportation Infrastructure Engineering. He has more than ten years of academic and professional experience at national and international levels, with a strong focus on risk assessment, safety analysis, and maintenance of road and airport infrastructures. His research activities include numerical modelling of pavement structural performance and surface characteristics, infrastructure risk evaluation, airport runway safety, and sustainable pavement maintenance strategies. He has participated in and coordinated research units within nationally funded research projects and has been involved in studies commissioned by qualified public and private stakeholders. He teaches courses in transportation infrastructure and has authored over 30 scientific publications and serves as Guest Editor and reviewer for several international scientific journals and conferences.

Preface

Road pavements represent a fundamental component of modern transportation infrastructure, directly influencing mobility, safety, economic development, and quality of life. As traffic demands increase and environmental constraints become more stringent, the design, construction, monitoring, and maintenance of road pavements face increasingly complex challenges. Addressing these challenges requires continuous innovation, interdisciplinary approaches, and the integration of advanced materials, modeling techniques, and monitoring technologies.

This reprint collects selected contributions from the Special Issue "Advances in Road Pavements" published in the journal Buildings (MDPI). The volume aims to provide a comprehensive overview of recent scientific and technological developments in road pavement engineering, highlighting both theoretical advancements and practical applications. The included chapters cover a wide range of topics, including innovative pavement materials and construction methods, pavement surface characteristics and safety, numerical modeling and simulation of pavement behavior, monitoring techniques, sustainability-oriented design approaches, and emerging challenges related to urban mobility.

The contributions reflect the diversity and dynamism of current pavement research, combining experimental investigations, numerical analyses, and real-world case studies. Particular emphasis is placed on improving pavement performance and durability while reducing environmental impacts and life-cycle costs. The reprint is intended to serve as a valuable reference for researchers, engineers, and practitioners involved in pavement engineering, as well as for policymakers and infrastructure managers seeking informed, science-based solutions.

By bringing together state-of-the-art research and applied studies, this volume aims to foster knowledge exchange and support the development of safer, more resilient, and more sustainable road pavements.

Emanuele Toraldo and Misagh Ketabdari
Guest Editors

buildings

MDPI

Article

Advanced Recycling of Modified EDPM Rubber in Bituminous Asphalt Paving

Daniela Laura Buruiana [1], Lucian Puiu Georgescu [2], Gabriel Bogdan Carp [1] and Viorica Ghisman [1,*]

[1] Interdisciplinary Research Centre in the Field of Eco-Nano Technology and Advance Materials CC-ITI, Faculty of Engineering, "Dunarea de Jos" University of Galati, 47 Domneasca, 800008 Galati, Romania; daniela.buruiana@ugal.ro (D.L.B.); gabriel.carp@ugal.ro (G.B.C.)

[2] European Center of Excellence for the Environment, Faculty of Sciences and Environment, "Dunarea de Jos" University of Galati, 800008 Galati, Romania; lucian.georgescu@ugal.ro

* Correspondence: viorica.ghisman@ugal.ro

Abstract: One of the environmental problems worldwide is the enormous number of surgical masks used during the COVID-19 pandemic due to the measures imposed by the World Health Organization on the mandatory use of masks in public spaces. The current study is a potential circular economy approach to recycling the surgical masks discarded into the environment during the COVID-19 pandemic for use in bituminous asphalt pavement. FTIR analysis showed that the surgical masks used were made from ethylene propylene diene monomer (EPDM) rubber modified with polypropylene. The effects of the addition of surgical masks in bituminous asphalt on the performance of the base course were demonstrated in this study. The morphology and elemental composition of the bituminous asphalt pavement samples with two ratios of surgical mask composition were investigated by SEM-EDX and the performance of the modified bituminous asphalt pavement was determined by Marshall stability, flow rate, solid–liquid ratio, apparent density, and water absorption. The study refers to the technological innovation of using surgical masks in the formulation of AB 31.5 bituminous asphalt base course, which brings tremendous benefits to the environment by reducing the damage caused by the COVID-19 pandemic.

Keywords: EDPM rubber; bituminous asphalt pavement; recycling

1. Introduction

The COVID-19 pandemic has resulted in significant health, financial, and environmental challenges worldwide [1]. The global public health initiative necessitated the use of surgical face masks as a crucial measure in combatting the spread of the coronavirus, leading to an increased demand for facemasks worldwide [2]. An estimated compound annual growth rate of 20% is projected for the supply of surgical and facial masks from 2020 to 2025, with this upward trend expected to persist [3]. The World Health Organization estimated at the beginning of the COVID-19-pandemic that approximately 89 million masks would be required each month to stem the spread of the pandemic [4]. Global production capacity of surgical masks is anticipated to rise alongside the surge in COVID-19 infection cases [5]. The effects of wearing facemasks on the cardiopulmonary system have been shown by several authors [6–10]. Disposable face masks, composed of polymeric materials, are entering the environment primarily through disposal in landfills, dumpsites, and littering in public spaces, subsequently finding their way into freshwater sources and oceans as a new source of microplastic fibers [11]. Currently, the primary recycling methods for disposable medical mask waste encompass high-temperature incineration, landfill decomposition, and chemical and mechanical recycling [12,13]. Surgical masks are obtained using different polymers and inorganic additives, and usually, there is no available detailed information on their composition [14], but the chemical composition of most disposable face masks is polypropylene, polyethylene, and polyurethane [15]. The high-temperature incineration method causes a variety of toxic byproducts that cause serious environmental pollution. The landfill degradation process involves breaking down polymers using soil microorganisms, but it is slow and can lead to secondary soil pollution. Mechanical recycling involves melting and combining pulverized disposable surgical masks with other materials to create lower-grade products. Chemical recycling converts high molecular weight polymers into smaller compounds through processes like pyrolysis or gasification, allowing them to be reformed into new materials. However, the current methods of recycling and treating waste surgical masks do not fully maximize waste utilization or generate significant energy and economic benefits for value-added utilization [12,13]. In order to efficiently minimize the impact of medical waste on humans and the environment, implementing a circular economy approach is crucial for success. Certain solid waste materials that have shown positive modification effects are being extensively utilized as modifiers in asphalt production [16]. To sustain the circular economy, we proposed in a previous study the use of recycled polypropylene and waste grit in a hot asphalt mixture used as the wear layer in road construction [17–19]. The authors of [20] conducted exploratory research on concrete incorporating face masks, demonstrating that the mechanical properties of concrete can be enhanced by integrating fibers made from surgical masks into the concrete mixture. Other researchers investigated the viability of incorporating disposable medical masks in asphalt binder, assessing its impact on the overall performance of asphalt. They observed positive effects on the high-temperature anti-rutting and low-temperature anti-cracking performance of both the asphalt binder and asphalt mixtures [21]. Researchers investigated the potential of repurposing face masks as an additive for hot mix asphalt to enhance the mechanical properties of asphalt pavement, creating a pathway to mitigate the rising pollution from personal protective equipment [22]. Other researchers stated that waste polypropylene fiber can significantly enhance the impact resistance and ductility of concrete, attributing this to its crack-bridging and reinforcing properties [23]. Several research studies in the literature have examined the fundamental mechanical properties of conventional aggregate concrete containing only face mask fibers [24]. The authors of one study devised a novel fiber-hybridization method to repurpose disposable medical face masks as fiber materials for creating environmentally friendly recycled concrete [25]. The used medical masks were repurposed as an agent to reduce viscosity and depress pour point in order to analyze its effectiveness in improving the flow characteristics of crude oil samples by authors [25]. The authors of further research presented a solution to use waste masks in fibered or crushed form in concrete. The results showed that both forms of mask waste can be used in concrete and the optimum value that increased the mechanical and durability properties was 0.5% crushed mask fiber [26]. Other researchers developed an expansion joint using styrene–butadiene–styrene as the filling material for rubber-modified asphalt which significantly improved the elasticity and adhesion of the asphalt mixture [27]. The authors of a later study designed a novel mask by modifying a surgical face mask with the adoption of a sealing frame and support and enhanced the filtration efficiency, concentration of CO_2, O_2, and N_2, dead space, and water vapor compared with the face masks from the market [28].

In order to overcome the issue of environmental pollution, this study proposed the use of surgical masks in bituminous asphalt pavement type AB 31.5. The investigation of bituminous asphalt pavement samples with two weight ratios of surgical masks examines the morphological characterization and the Marshall characteristics. The novelty of incorporating surgical masks into bituminous asphalt paving, particularly as a base layer in road construction, lies in the potential to address environmental concerns effectively. Despite existing studies on this topic, the focus on minimizing environmental impact by repurposing surgical masks in road construction can lead to significant sustainability benefits. This innovative approach aims to offer a practical solution for reusing medical waste while also enhancing the performance and durability of road infrastructure.

2. Materials and Methods

2.1. Life Cycle Assessment

The goal of this study is to recycle surgical masks in bituminous asphalt pavement with the aim of extending their life cycle and implementing circular economy principles for responsible waste management. In Figure 1, the life cycle assessment of surgical masks with a sustainable approach to extending their life cycle is presented.

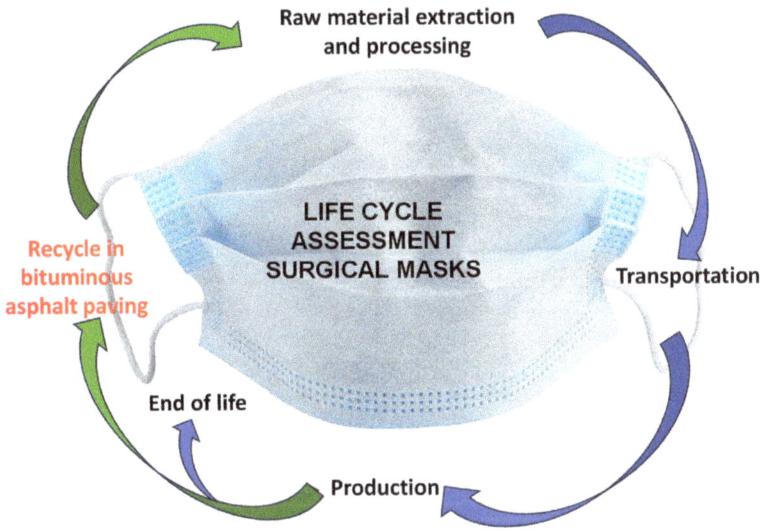

Figure 1. Life cycle assessment of surgical masks.

Implementing the circular economy by extending the life cycle of protective equipment such as surgical masks for use in technical-grade road construction is an innovative and sustainable approach. Extending the life cycle of surgical masks by recycling and using them in the construction of road base layers can bring several benefits.

The use of recycled materials contributes to reducing the consumption of natural resources and minimizing the waste generated, thereby having a positive impact on the environment. Surgical masks can find a new useful life without being thrown into the environment and polluting the soil or water. Such practices support the circular economy, promoting the reuse and recycling of materials, reducing carbon emissions, and contributing to efficient resource management. The use of surgical masks in road construction can bring practical benefits, such as improving insulation properties, water resistance, and the ability to stabilize the soil in the road base layer. This practice can lead to reduced construction and maintenance costs of road infrastructure by using more durable and sustainable materials. Thus, the implementation of the circular economy in extending the life cycle of surgical masks for their use in road infrastructure can bring advantages both from an environmental and economic point of view, contributing to more efficient management of resources and protecting the environment for future generations.

2.2. Bituminous Asphalt Paving Sample Preparation

The hot asphalt mix used in this study consisted of base courses with the designation bituminous asphalt pavement AB 31.5, whereby the value 31.5 indicates the maximum grain size of the granulate [29]. The samples of bituminous asphalt pavement type AB 31.5 were prepared in accordance with the indicative normative AND 546-2013, Norm on the hot execution of bituminous coatings for the bridge path, (2013) and AND 605-2016: Normative regarding hot asphalt mixes [29,30], tested in the Laboratory for Analysis and

Testing in Construction—Class II of SC. Antrepriza de Constructii Drumuri si Autostrazi SRL Road Company. The weight percentages of the components of the standard base course of bituminous asphalt type AB 31.5 consist of 40.8% natural screened aggregate with a size greater than 4.0 mm, 50% crushed sand with a grain size between 0.1 and 4.0 mm, 5% limestone filler, and 4.2% road bitumen type 50/70 [29,31]. The bitumen properties of penetration at 25 °C type 50/70 are as follows: softening point 46/54 °C; resistance to hardening at 163 °C: change in mass 0.5%, retained penetration 50 min, softening point after hardening 58 °C, flashpoint 230 min; solubility at specification 99 min; kinematic Viscosity 135 °C specification 295 min [32]. Two formulations were prepared for the Marshall test with the addition of a percent by weight of surgical masks. Surgical masks used (worn surgical masks cannot be used in the laboratory due to public safety measures for the prevention of SARS-CoV-2 contamination). When placed in the casting mixture (160 °C), there is no risk of soil contamination/population and, due to the fact that the sterilization process takes place at a temperature of 130 °C, the masks become inert. Thus, the weight proportions of the asphalt mix sample constituents consist of 40.8% natural screened aggregate with a size above 4.0 mm, 50% crushed sand with a particle size between 0.0 and 4.0 mm, 5% graded limestone filler with a particle size of 0.063 and 0.100 mm, 4.1% paving bitumen type 50/70 with 0.1% surgical masks for sample 1, and 3.9% paving bitumen type 50/70 with 0.3% surgical masks for sample 2, as shown in Table 1. The bituminous asphalt paving samples were modified with the masks by bitumen replacement, i.e., a percent of 0.1% surgical masks added for Sample 1 and a percent of 0.3% surgical masks added for Sample 2, with a bitumen percent of 4.1% for Sample 1 and 3.9% for Sample 2.

The formulation of hot bituminous asphalt type AB 31.5 containing 0.3% surgical masks as used in Sample 2 was the subject of the patent filed in 2021 [33]. The samples used for Marshall tests have a cylindrical shape with a diameter of 10 cm and a height of 6.3 cm.

Figure 2 shows the gradation determined for the bituminous asphalt paving, and the size distribution of the aggregate particles used in the asphalt composition for the standard sample and for the sample with a surgical mask content of 0.3% can be seen. The grading results show a well-balanced granulometry that ensures a homogeneous mixture, offering resistance and long-term durability.

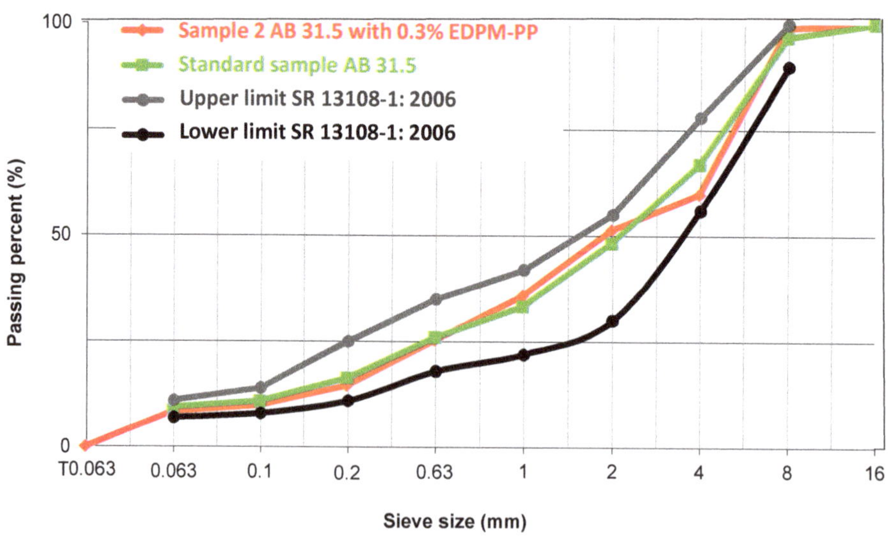

Figure 2. Gradation of standard bituminous asphalt pavement type AB 31.5 and Sample 2 (0.3% EDPM-PP).

Table 1. Recipes for bituminous asphalt pavement samples type AB 31.5.

Component	Standard	Sample 1	Sample 2
Crushed siliceous stone [%]	40.8	40.8	40.8
Crushed sand [%]	50	50	50
Sort limestone filler [%]	5	5	5
Road bitumen 50/70 [%]	4.2	4.1	3.9
Surgical masks [%]	-	0.1	0.3

2.3. Characterization Methods

The quantitative determination was conducted using the FTIR-attenuated total reflection (FTIR-ATR) method, which records spectra through attenuated total reflection. The Fourier transform infrared (FTIR) spectra of the surgical mask were captured using an IR-Spirit-T FTIR Spectrometer from Shimadzu, equipped with a built-in ATR accessory type QATR-S, DLATGS detector, and KBr beam splitter. The scan range was set between 400–4600 cm^{-1} with a resolution of 2 cm^{-1}, and each scan was repeated 45 times. The FTIR spectrometer was operated within a temperature-controlled (20 °C) air-conditioned chamber.

The morphology and elemental composition of the bituminous asphalt paving samples were examined by scanning electron microscopy (SEM) in high vacuum conditions with a 4th generation TESCAN VEGA electron source and tungsten filament which combines SEM imaging and elemental composition analysis directly in a single Essence™ from TESCAN software (1.2.1.0 build 5762, March 2023) window. This integration greatly streamlines the collection of morphological and elemental information from the sample, positioning VEGA SEM as a highly effective analytical tool for routine material evaluation in quality control, failure analysis, and research settings. Prior to analysis, the samples underwent a coating process with a 6 nm thick conductive layer of Au utilizing the SPI-Module™ sputter coater system.

The physical and mechanical properties were determined through tests conducted on Marshall cylinders. The test principle involves identifying the compressive strength of a cylindrical sample when subjected to a force applied by a generator. This test is performed on a sample contained within the mold at a temperature of 60 °C. The Marshall method is a common approach used to determine the optimal mix design for asphalt paving. The Marshall method provides a systematic approach to designing high-quality asphalt mixes that meet the specific requirements of a paving standard. The general steps involved for the Marshall method are as follows: selection of aggregate based on factors such as size, shape, and gradation to ensure good compaction and durability of the asphalt mix; selection of asphalt binder; mixing of design through blending of components in the right proportions; compaction testing of the bituminous asphalt paving in a mold to a specific density using a hammer; and density and stability testing. The compacted samples are then subjected to testing to determine parameters such as density, stability, flow, and solid–liquid ratio.

The physical–mechanical characterizations were performed according to Romanian Standard SR EN 12697-6 [34] and SR EN 12697-34 [35] as follows:

Stability (S) represents the load at which the cylindrical specimen fractures at 60 °C, measured in kN.

The flow index (I), measured in mm, indicates the deformation of the vertical diameter of the specimen at the point of fracture.

The apparent density (g/cm^3) is the mass of a unit volume of compacted bituminous asphalt pavement, accounting for the air-filled voids. It is calculated using the formula that relates sample mass to volume (1):

$$\rho_a = \frac{m_u}{V} \left(\frac{g}{cm^3}\right) \quad (1)$$

For the standard sample, the following holds:

- ρ_a represents the apparent density of the bituminous asphalt pavement (g/cm^3);
- m_u is the mass of the sample measured on the analytical balance in a dry state;
- V denotes the volume of the cylindrical sample with a diameter of 10 cm and a height of 6.3 cm.

Water absorption [%] refers to the amount of water absorbed by the externally accessible voids in a bituminous asphalt paving sample. This was determined through evaporation after immersing the sample in water using the static thermal method. Following removal from the water, the sample was wiped with a damp cloth to eliminate excess surface and gravity water. Subsequently, the wet sample underwent drying in an oven at 60 °C and relative humidity below 2.5%. The water absorption or wetting capacity is expressed as a percentage of the initial sample mass and the mass after wetting. The volume percentage of water absorption is calculated using the following relationship:

$$A = 100\frac{m_i}{m_f} \tag{2}$$

where the following holds:

- m_i represents the initial mass of the sample.
- m_f is the final mass of the sample after wetting.

Each test result in this research study is the average of three replicate tests.

3. Results and Discussion

3.1. Structural Characterization of Bituminous Asphalt Paving Samples

The resulting FTIR spectra of components of the surgical mask used in this study are shown in Figure 3. The surgical mask is made of an elastic ear loop (blue color), heat-welded seam (brown color), and non-woven fabric (red color). Analyzing the FTIR database revealed that the tested surgical mask exhibited similarities with the spectrum of ethylene propylene diene (EPDM) rubber modified with polypropylene (PP). The FTIR spectrum showed intense bands in the wavenumber 2914, 2354, 1643, 1454, 1372, 1161, 827, 670 and 448 cm^{-1}.

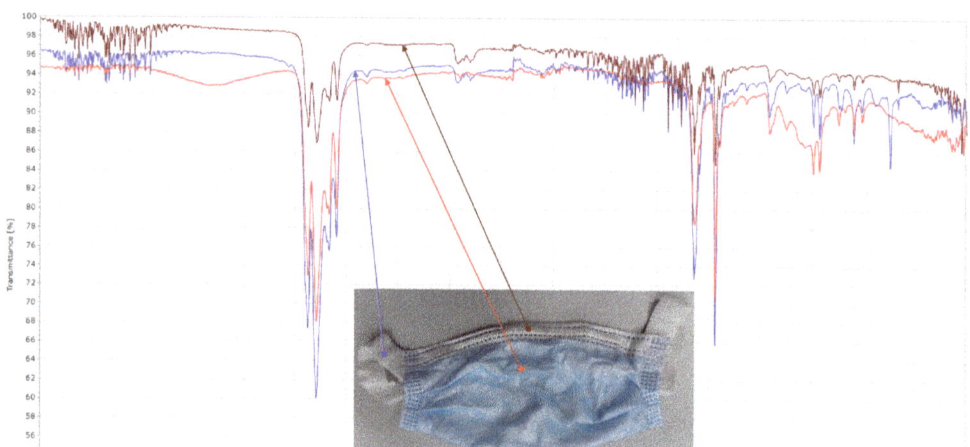

Figure 3. FTIR spectra of surgical mask.

The transmittance band distribution on the FTIR spectra of the surgical mask sample showed broad ranges from 1161 to 1454 cm^{-1} and 2914 cm^{-1} attributed to CH_2 asymmetric stretching, which are characteristic for the FTIR spectrum of polypropylene (PP) [36]. The FTIR spectrum also showed bands at 1643, 827, 670, and 448 cm^{-1}; the band at 1643 cm^{-1} was attributed to C=C stretching, which is characteristic for EPDM, the weak transmittance band at 827 cm^{-1} could be attributed to the possible degradation (cross-linking) processes between EPDM rubber and PP, and the band at 670 cm^{-1} corresponds to the Si-O-Mg bond in talc filler [37], while the band from 448 cm^{-1} is associated with the S-S of cross-linked EPDM-PP [38]. The FTIR spectra of the surgical masks used in this study showed that the chemical composition is mainly ethylene propylene diene (EPDM) rubber modified with polypropylene (PP). In Figure 4 are presented the FTIR spectrum of spectra of standard bituminous asphalt pavement (black), Sample 1 (0.1% EDPM-PP) (green), and Sample 2 (0.3% EDPM-PP) (orange). The FTIR spectra of all three components of the surgical mask display almost identical peaks, albeit with varying intensities, distinctly indicating the presence of EPDM-PP.

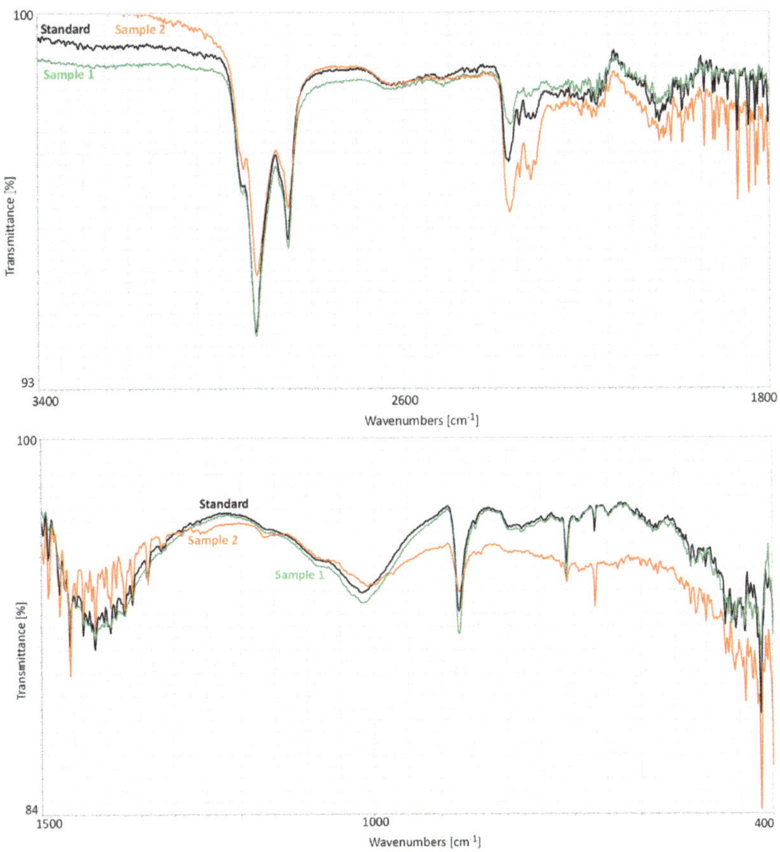

Figure 4. FTIR spectra of standard bituminous asphalt pavement (black), Sample 1 (0.1% EDPM-PP) (green), and Sample 2 (0.3% EDPM-PP) (orange).

Figure 5 depicts SEM images of the standard bituminous asphalt pavement type AB 31.5, Sample 1, and Sample 2. The standard sample appears smooth and uniform. Observably, a compact arrangement between the components is evident with the addition of EPDM-PP in Sample 1 and Sample 2. Due to the different surface microstructures of the bituminous asphalt pavement, the structure of the interface of the standard sample was different from that of the samples with EPDM-PP (Sample 1 and Sample 2). The morphology of Sample 1 shows a higher porosity than the sample with higher EPDM-PP content.

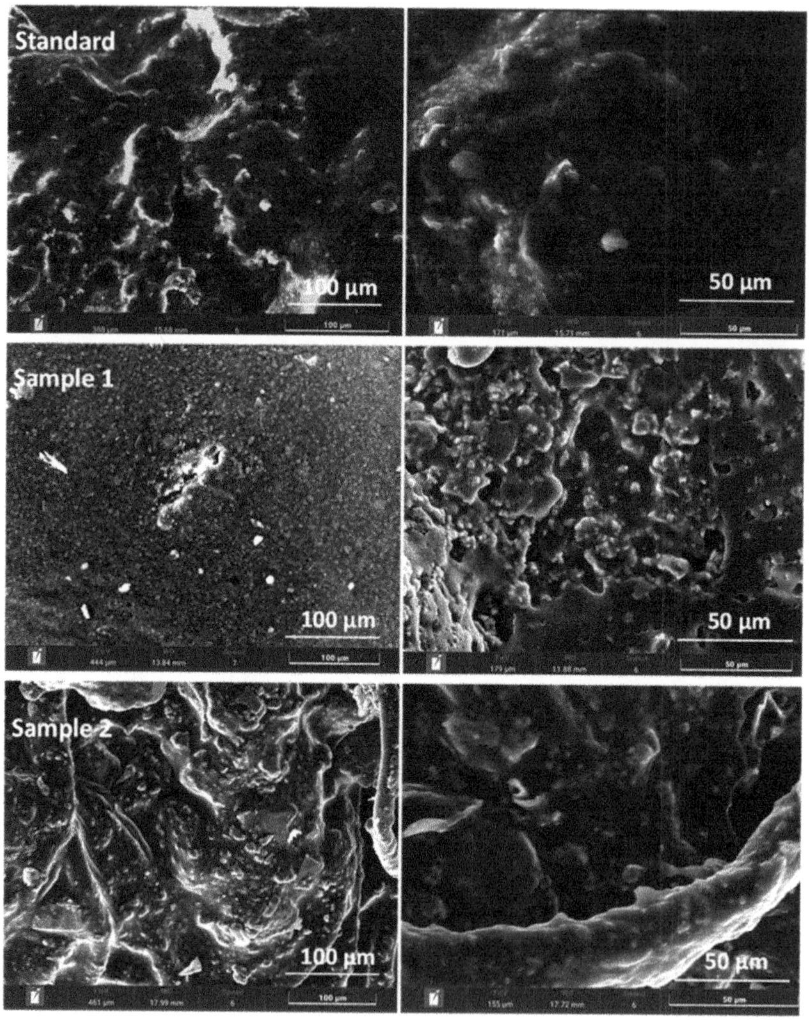

Figure 5. SEM images of standard bituminous asphalt pavement, Sample 1 (0.1% EDPM-PP), and Sample 2 (0.3% EDPM-PP).

In the SEM image of Sample 2 (containing 0.3% EPDM-PP), it is evident that the crushed silica crisps, crushed sand, and limestone filler are effectively embedded in a polymer matrix. This embedding matrix and the interconnections between the components exhibit minimal pores and reduced roughness.

After the asphalt was embedded in the polymer matrix, its loading area increased due to the content of surgical masks, which significantly improved its resistance to external forces. In Sample 2, the fibers originating from the surgical masks that connect the components of the asphalt mix can be seen at higher magnification.

The elemental composition of Sample 2 (with 0.3% surgical masks), consisting of the most important chemical elements C, O, Ca, and Si, is shown in Figure 6. The high concentration of carbon, oxygen, and silicon indicates the presence of surgical masks. It can be noted that the fibrous structures have a high percentage of carbon, most likely originating from polypropylene staple fibers. The presence of calcium (Ca) is also favorable due to its good interaction with bitumen.

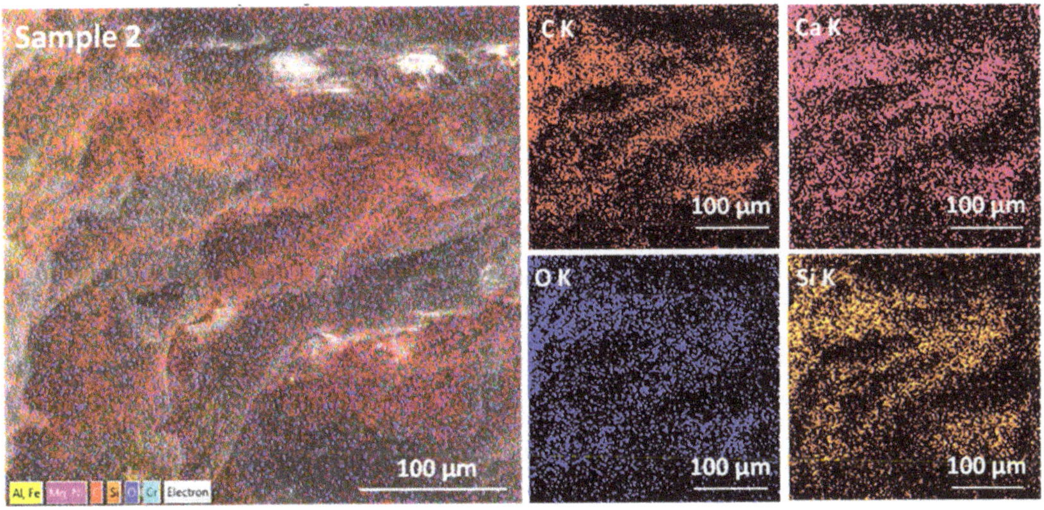

Figure 6. EDX elemental map of Sample 2 (0.3% EDPM-PP) bituminous asphalt pavement.

3.2. Physical–Mechanical Properties of Bituminous Asphalt Paving Samples

The Marshall test is a crucial laboratory procedure used to evaluate the properties of bituminous asphalt pavement, including those improved with EDPM-PP (ethylene propylene diene monomer–polypropylene) additives. This test provides valuable insights into the performance characteristics of asphalt mixes, helping to ensure that they meet specified requirements for use in road construction and paving projects. Here are some key aspects of the Marshall test as it pertains to bituminous asphalt pavement improved with EDPM-PP:

The primary objective of the Marshall test for bituminous asphalt pavement improved with EDPM-PP is to assess its mechanical properties, durability, and resistance to deformation under simulated traffic and environmental conditions. By subjecting the asphalt specimens to controlled loading and temperature conditions, the test aims to determine their stability, flow, density, and other key performance indicators.

During the sample preparation phase, the bituminous asphalt pavement improved with EDPM-PP is compacted in the Marshall molds to achieve the desired density and air void content. The compaction process involves applying a specific number of blows with a compaction hammer to ensure proper aggregate particle orientation and interlocking, as well as consistent distribution of the bitumen and EDPM-PP additives within the mixture. Once the specimens are prepared and conditioned, they are subjected to stability and flow tests under controlled loading and temperature conditions. The stability test measures the resistance of the bituminous asphalt pavement to deformation and rutting, providing insights into its ability to withstand traffic-related stresses. The flow test, on the other hand, assesses the deformation or consolidation of the specimen under compressive loading, indicating its flow characteristics under applied pressure.

In addition to stability and flow testing, the Marshall test involves evaluating the compacted density and air void content of the asphalt specimens improved with EDPM-PP. These analyses provide critical information about the degree of compaction achieved during specimen molding, as well as the amount of air voids present in the bituminous asphalt pavement, which can impact its long-term performance, permeability, and resistance to moisture damage.

The results of the Marshall test for bituminous asphalt paving improved with EDPM-PP play a critical role in predicting the performance of the bituminous asphalt pavement under actual field conditions. Additionally, the test data can be used to optimize the mix design by adjusting the composition, proportion of additives, and binder content to achieve the desired performance targets, such as enhanced resistance to cracking, fatigue, and deformation.

In the context of regulatory compliance and quality assurance, the Marshall test serves as a tool for verifying that the EDPM-PP-modified bituminous asphalt pavement meets the specified performance requirements and industry standards. Compliance with test-based performance criteria is essential for ensuring the durability, safety, and longevity of asphalt pavements, particularly when enhanced with additives such as EDPM-PP.

Marshall stability measures the maximum load that bituminous asphalt can withstand at a temperature of 60 °C and correlates well with measurements of rutting in bituminous asphalt pavement during operation.

In line with STAS recommendations [34,35], AB 31.5 bituminous asphalt pavement should ideally exhibit a Marshall stability of at least 6.3 kN. All test results of the bituminous asphalt samples meet the performance standards in Romania, as depicted in Figure 4. Notably, the rut resistance improves with the rising EDPM-PP content. With an increase in EDPM-PP content to 0.3% (as seen in Sample 2), the mixture's strength enhances due to the presence of polymer matrix fibers, impacting gradation and mechanical properties. Figure 4 illustrates that stability increases with higher surgical mask content, attributed to its lower specific gravity compared to standard asphalt. This allows it to penetrate between particles, enhancing aggregate interlocking and overall stability. Consequently, the bituminous asphalt featuring 0.3% EDPM-PP (Sample 2) exhibits optimal stability, leading to improved deformation resistance in the mixture. The flow rate, as per STAS 12697-34 [35], can range between 1.5 to 4.5 mm. In Figure 7, the flow rates of both standard and modified hot asphalt mix samples are illustrated. It is evident from the figure that the flow rate rises with the inclusion of EDPM-PP but decreases notably at a higher content of 0.3% EDPM-PP. This shift can be correlated with the enhanced interlocking of fibers from the surgical masks with the asphalt binder and aggregates. The flow values escalate with the surgical mask content, suggesting increased deformations under the same pressure. This observation underscores that as the EDPM-PP content elevates, the resistance to deformation improves, particularly when the surgical mask content reaches 0.3%. Also, the flow rate values of bituminous asphalt pavement samples meet the performance standards needed in Romania for this type of bituminous asphalt pavement.

The solid–liquid ratio must meet a minimum threshold of 1.6 kN/mm. As illustrated in Figure 8, both standard and bituminous asphalt pavement samples containing EDPM-PP adhere to Romania's performance standards. Notably, the solid–liquid ratio sees an increase with a higher EDPM-PP content, indicative of the strengthening effect as the surgical masks interacts with bitumen. Additionally, Figure 8 displays the water absorption levels for both the standard and modified bituminous asphalt pavement samples featuring EDPM-PP. Per Romanian Standard STAS 12697-34/2020, water absorption values should fall within the range of 1.5–6.0%. In samples with a minimal percentage of surgical masks, water absorption aligns with the standard sample, decreasing as the EDPM-PP content rises.

Furthermore, Figure 9 depicts the apparent density of bituminous asphalt pavement samples. For AB 31.5 bituminous asphalt pavement, the apparent density should not be less than 2.330 g/cm^3 in accordance with SR 12697-6/2020.

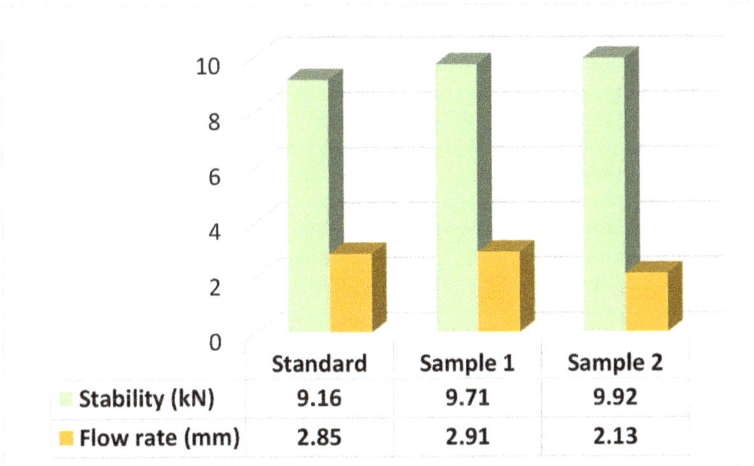

Figure 7. Stability at 60 °C and flow rate of standard bituminous asphalt pavement, Sample 1 (0.1% EDPM-PP), and Sample 2 (0.3% EDPM-PP).

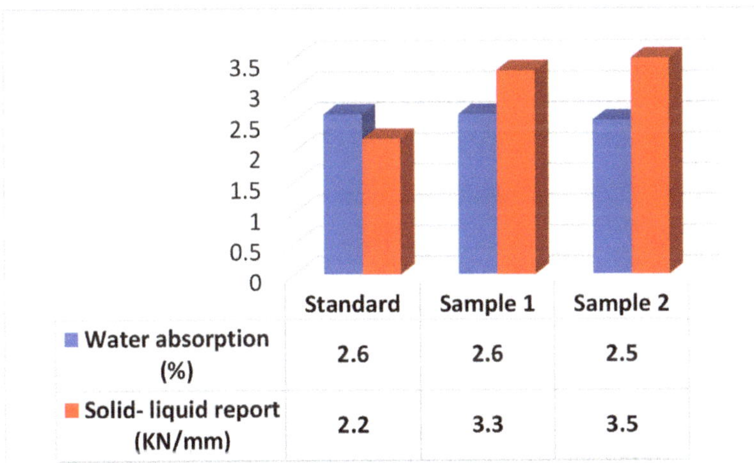

Figure 8. Solid–liquid report and water absorption of standard bituminous asphalt pavement, Sample 1 (0.1% EDPM-PP), and Sample 2 (0.3% EDPM-PP).

All bituminous asphalt pavement samples comply with Romania's performance standards. The solid–liquid ratio values decline with higher EDPM-PP content. The base layer's performance is closely linked to the density of the bituminous asphalt pavement. It is evident that the apparent density increases with the addition of surgical masks. Achieving a higher apparent density is crucial to prevent potential rutting issues caused by insufficient air voids due to base layer densification from traffic loads. In summary, the sample containing 0.3% surgical masks recorded the highest apparent density among the samples tested.

Incorporating EDPM-PP derived from surgical masks into bituminous asphalt pavement used as a base layer in road construction can have significant environmental benefits. Despite the existing research findings, there are several reasons to consider this approach from an environmental standpoint: utilizing EDPM-PP from surgical masks in asphalt pavement provides a sustainable solution for recycling these materials, reducing waste in landfills, and promoting a circular economy approach; by repurposing EDPM-PP in road construction, we reduce the dependency on virgin materials, conserving natural resources and lowering the environmental impact associated with traditional construction materials; incorporating recycled materials like EDPM-PP can lead to energy savings during production processes, as recycling typically requires less energy compared to manufacturing new materials from raw sources; recycling EDPM-PP from surgical masks in asphalt pavement helps mitigate greenhouse gas emissions associated with conventional production methods, contributing to overall carbon footprint reduction; reusing EDPM-PP in road construction helps alleviate environmental pollution by preventing these materials from ending up in landfills or being improperly disposed of, thus reducing the burden on ecosystems and waste management systems; and embracing the use of recycled materials like EDPM-PP demonstrates a commitment to sustainable practices in construction, aligning with broader environmental objectives and fostering a culture of responsible resource management.

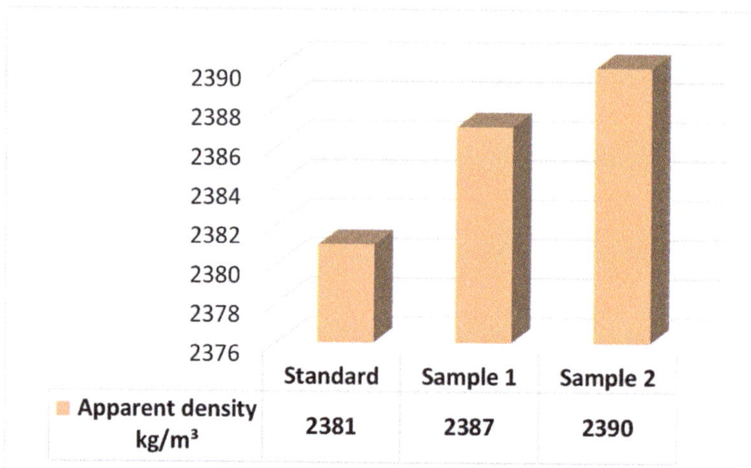

Figure 9. Apparent density of standard bituminous asphalt pavement, Sample 1 (0.1% EDPM-PP), and Sample 2 (0.3% EDPM-PP).

Incorporating EDPM-PP from surgical masks in bituminous asphalt pavement not only offers environmental benefits but also supports the transition towards more eco-friendly infrastructure development, contributing to a greener and more sustainable future.

4. Conclusions

One solution to address pressing environmental concerns stemming from the increased use of surgical masks during the COVID-19 pandemic, in line with World Health Organization directives mandating their use in public spaces, is recycling to mitigate environmental impact. This study presents an innovative approach to combat pandemic-generated waste by repurposing used face masks in bituminous asphalt pavement. Analyses utilizing FTIR spectroscopy revealed that the composition of the surgical masks studied closely resembles ethylene propylene diene (EPDM) rubber modified with polypropylene (PP). Incorporating EDPM-PP into the base layer mix of bituminous asphalt pavement type AB 31.5 resulted in enhanced performance of the mixture, as confirmed by test results.

SEM analysis indicated that when 0.3% EDPM-PP was added to the bituminous asphalt pavement, the crushed siliceous stone chipping, crushed sand, and limestone filler were effectively integrated into a polymer-matrix-like structure, exhibiting increased resistance to external forces. The study observed improvements in the Marshall properties of the bituminous asphalt pavement with the inclusion of ethylene propylene diene rubber modified with polypropylene, wherein higher stability, solid–liquid ratio, and apparent density were noted, attributed to the fibrous reinforcement from the surgical masks enhancing aggregate interlock. Following analysis of Marshall results, 0.3 percent EDPM-PP was identified as the optimal content for the mixture.

In summary, the utilization of surgical masks as a cohesive material proved effective in stabilizing bituminous asphalt pavement components, thereby mitigating the significant environmental impacts resulting from the COVID-19 pandemic on a global scale. The findings of this research present valuable insights for industry stakeholders and government bodies seeking to explore sustainable practices within the pavement sector. Overall, the implementation of circular economy strategies in extending the lifespan of surgical masks through their incorporation into road infrastructure showcases a win-win scenario. It not only contributes to a more sustainable and environmentally friendly approach, but also reaps economic benefits by optimizing resource utilization and paving the way for a greener future for generations to come. Future studies aim to further experiment with innovative approaches to advance sustainable solutions in road construction.

5. Patents

The recipe of Sample 2 of bituminous asphalt pavement type AB 31.5 with a content of 0.3% of surgical masks was the subject of the patent entitled RECYCLING OF SURGICAL MASKS IN HOT ASPHALT MIXTURES, RO135384A0, 2021.

Author Contributions: Conceptualization, D.L.B., G.B.C., and V.G.; methodology, D.L.B., L.P.G., and V.G.; formal analysis and investigation, L.P.G. and G.B.C.; writing—original draft preparation, D.L.B. and V.G. All authors have read and agreed to the published version of the manuscript.

Funding: This research received no external funding.

Institutional Review Board Statement: Not applicable.

Data Availability Statement: Data are contained within the article.

Acknowledgments: The authors would like to acknowledge the support received of the project "Excellence and involvement in intelligent development based on research and innovation at the "Dunarea de Jos" University of Galati-DINAMIC", ID 536/2021. The authors thanks to the Tatiana Mardare from Urban Office Arhitecture, Milano, Italia for support and Grade II from S C. Antrepriza de Constructii Drumuri si Autostrazi SRL Road Company for laboratory Marshall tests of bituminous asphalt pavement samples.

Conflicts of Interest: The authors declare no conflicts of interest.

References

1. Garel, A.; Petit-Romec, A. Investor rewards to environmental responsibility: Evidence from the COVID-19 crisis. *J. Corp. Financ.* **2020**, *68*, 101948. [CrossRef]
2. Royo-Bordonada, M.A.; García-López, F.J.; Cortés, F.; Zaragoza, G.A. Face masks in the general healthy population. Scientific and ethical issues. *Gac. Sanit.* **2021**, *35*, 580–584. [CrossRef] [PubMed]
3. Ilyas, S.; Srivastava, R.R.; Kim, H. Disinfection technology and strategies for COVID-19 hospital and bio-medical waste management. *Sci. Total Environ.* **2020**, *749*, 141652. [CrossRef] [PubMed]
4. WHO. *Shortage of Personal Protective Equipment Endangering Health Workers Worldwide*; WHO: Geneva, Switzerland, 2020.
5. Phan, T.L.; Ching, C.-S. A Reusable Mask for Coronavirus Disease 2019 (COVID-19). *Arch. Med. Res.* **2020**, *51*, 455–457. [CrossRef] [PubMed]
6. Guan, T.; Hu, S.; Han, Y.; Wang, R.; Zhu, Q.; Hu, Y.; Fan, H.; Zhu, T. The effects of facemasks on airway inflammation and endothelial dysfunction in healthy young adults: A double-blind, randomized, controlled crossover study. *Part. Fibre Toxicol.* **2018**, *15*, 30. [CrossRef] [PubMed]
7. Bobeica, C.; Niculet, E.; Tatu, A.L.; Craescu, M.; Vata, D.; Statescu, L.; Iancu, A.V.; Musat, C.L.; Draganescu, M.L.; Onisor, C.; et al. Old and new therapeutic strategies in systemic sclerosis (Review). *Exp. Ther. Med.* **2022**, *23*, 134. [CrossRef] [PubMed]
8. Bobeica, C.; Niculet, E.; Craescu, M.; Parapiru, E.L.; Musat, C.L.; Dinu, C.; Chiscop, I.; Nechita, L.; Debita, M.; Stefanescu, V.; et al. CREST Syndrome in Systemic Sclerosis Patients—Is Dystrophic Calcinosis a Key Element to a Positive Diagnosis? *J. Inflamm. Res.* **2022**, *15*, 3387–3394. [CrossRef] [PubMed]

9. Schmidt, N.; Thibault, D.; Galgani, F.; Paluselli, A.; Sempéré, R. Occurrence of microplastics in surface waters of the Gulf of Lion (NW Mediterranean Sea). *Prog. Oceanogr.* **2018**, *163*, 214–220.
10. Chen, H.; Dong, X.; Zhao, Y.; Wang, D. Recycling and chemical upgrading of discarded disposable medical masks. *Acta Polym. Sin.* **2020**, *51*, 12.
11. Dharmaraj, S.; Ashokkumar, V.; Hariharan, S.; Manibharathi, A.; Ngamcharussrivichai, C.J.C. The COVID-19 pandemic face mask waste: A blooming threat to the marine environment. *Chemosphere* **2021**, *272*, 129601. [CrossRef] [PubMed]
12. Jung, S.; Lee, S.; Dou, X.; Kwon, E.E. Valorization of disposable COVID-19 mask through the thermo-chemical process. *Chem. Eng. J.* **2021**, *405*, 126658. [CrossRef] [PubMed]
13. Yesil, Y.; Bhat, G.S. Structure and mechanical properties of polyethylene melt blown nonwovens. *Int. J. Cloth. Sci. Technol.* **2016**, *28*, 780–793. [CrossRef]
14. Wang, S.; Gao, Y.; Yan, K.; You, L.; Jia, Y.; Dai, X.; Chen, M.; Diab, A. Effect of long-term aging on waste tire rubber and amorphous poly alpha olefin compound modified asphalt binder and its mixtures. *Constr. Build. Mater.* **2021**, *272*, 121667. [CrossRef]
15. Buruiana, D.L.; Georgescu, P.L.; Carp, G.B.; Ghisman, V. Recycling micro polypropylene in modified hot asphalt mixture. *Sci. Rep.* **2023**, *13*, 3639. [CrossRef] [PubMed]
16. Buruiana, D.L.; Bordei, M.; Sandu, A.V.; Chirculescu, A.I.; Sandu, I.G. Studies on Grit Use in Asphalt Mixtures (II). *Mat. Plast.* **2013**, *50*, 113–118.
17. Buruiana, D.L.; Bordei, M.; Sandu, I.G.; Chirculescu, A.I.; Sandu, I. Recycling Waste Grit in Mix Asphalt. *Mat. Plast.* **2013**, *50*, 36–39.
18. Kilmartin-Lynch, S.; Saberian, M.; Li, J.; Roychand, R.; Zhang, G. Preliminary evaluation of the feasibility of using poly-propylene fibres from COVID-19 single use face masks to improve the mechanical properties of concrete. *J. Clean. Prod.* **2021**, *296*, 126460. [CrossRef] [PubMed]
19. Zhao, Z.; Wu, S.; Liu, Q.; Xie, J.; Yang, C.; Wang, F.; Wan, P. Recycling waste disposable medical masks in improving the performance of asphalt and asphalt mixtures. *Constr. Build. Mater.* **2022**, *337*, 127621. [CrossRef]
20. Wang, G.; Li, J.; Saberian, M.; Rahat, H.H.; Massarra, C.; Buckhalter, C.; Farrington, J.; Collins, T.; Johnson, J. Use of COVID-19 single-use face masks to improve the rutting resistance of asphalt pavement. *Sci. Total Environ.* **2022**, *826*, 154118. [CrossRef] [PubMed]
21. Alrshoudi, F.; Mohammadhosseini, H.; Tahir, M.M.; Alyousef, R.; Alghamdi, H.; Alharbi, Y.R.; Alsaif, A. Sustainable Use of Waste Polypropylene Fibers and Palm Oil Fuel Ash in the Production of Novel Prepacked Aggregate Fiber-Reinforced Concrete. *Sustainability* **2020**, *12*, 4871. [CrossRef]
22. Koniorczyk, M.; Bednarska, D.; Masek, A.; Cichosz, S. Performance of concrete containing recycled masks used for personal protection during coronavirus pandemic. *Constr. Build. Mater.* **2022**, *324*, 126712. [CrossRef] [PubMed]
23. Boarie, A.; Abdelsalam, M.; Gamal, A.; Rabah, M. Laboratory and Environmental Assessment of Asphalt Mixture Modified with a Compound of Reclaimed Asphalt Pavement and Waste Polyethylene. *Buildings* **2024**, *14*, 1186. [CrossRef]
24. Ahmed, W.; Lim, C.W. Effective recycling of disposable medical face masks for sustainable green concrete via a new fiber hybridization technique. *Constr. Build. Mater.* **2022**, *344*, 128245. [CrossRef] [PubMed]
25. Wang, P.; Gu, X.; Xue, M.; Li, Y.; Dong, S.; Chen, G.; Zhang, J. Resource utilization of medical waste under COVID-19: Waste mask used as crude oil fluidity improver. *J. Clean. Prod.* **2022**, *358*, 131903. [CrossRef] [PubMed]
26. Idrees, M.; Akbar, A.; Mohamed, A.M.; Fathi, D.; Saeed, F. Recycling of Waste Facial Masks as a Construction Material, a Step towards Sustainability. *Materials* **2022**, *15*, 1810. [CrossRef] [PubMed]
27. Kim, K.-N.; Kim, Y.-M.; Le, T.H.M. Development of Plug Joint with Polymer-Modified Rubber Asphalt as Filling Material. *Polymers* **2023**, *15*, 4256. [CrossRef] [PubMed]
28. Zhang, T.; Zhang, T.; Liu, S. A Modified Surgical Face Mask to Improve Protection and Wearing Comfort. *Buildings* **2022**, *12*, 663. [CrossRef]
29. *STAS SR EN 13108-1: 2006*; Asphalt Mixtures. Specifications for Materials. Part 1: Asphalt Concrete. National Road Infrastructure Administration Company: Brussels, Belgium, 2006.
30. *AND 546-2013*; Norm on the Hot Execution of Bituminous Coatings for the Bridge Path. National Road Infrastructure Administration Company: Brussels, Belgium, 2013.
31. *AND 605-2016;* Normative Regarding Hot Asphalt Mixes. Technical Conditions for Design, Preparation and Putting into Operation of Asphalt Mixtures. National Road Infrastructure Administration Company: Brussels, Belgium, 2016.
32. *European Standard DIN EN 12591:99*; Technical considerations regarding the quality of road bitumen. National Road Infrastructure Administration Company: Brussels, Belgium, 1999.
33. Buruiana, D.L.; Georgescu, P.L.; Carp, G.B.; Ghisman, V.; Mardare, T. Recycling of Surgical Masks in Hot Asphalt Mixtures, Nr. A 2021 00468, RO135384A0. 2021. Available online: https://www.webofscience.com/wos/diidw/full-record/DIIDW:2022166774 (accessed on 30 January 2024).
34. *SR EN 12697-6:2020*; Asphalt Mixtures. Test Methods. Part 6: Determination of Bulk Density of Bituminous Samples. National Road Infrastructure Administration Company: Brussels, Belgium, 2020.
35. *SR EN 12697-34:2020*; Asphalt Mixtures. Test Methods. Part 34: Marshall Test. National Road Infrastructure Administration Company: Brussels, Belgium, 2020.

36. Jung, M.R.; Horgen, F.D.; Orski, S.V.C.V.R.; Beers, K.L.; Balazs, G.H.; Jones, T.T.; Work, T.M.; Brignac, K.C.; Royer, S.J.; Hyrenbach, K.D.; et al. Validation of ATR FT-IR to identify polymers of plastic marine debris, including those ingested by marine organisms. *Mar. Pollut. Bull.* **2018**, *127*, 704–716. [CrossRef] [PubMed]
37. Ossman, M.; Mansour, M.; Fattah, M.; Taha, N.; Kiros, Y. Peanut shells and talc powder for removal of hexavalent chromium from aqueous solutions. *Bulg. Chem. Commun.* **2014**, *46*, 629–639.
38. Sarfraz, M.; Rehman, Z.; Ba-Shammakh, M. Pursuit of electroconducting thermoplastic vulcanizates: Activated charcoal-filled polypropylene/ethylene–propylene–diene monomer blends with upgraded electrical, mechanical and thermal properties. *Polym. Bull.* **2019**, *76*, 2005–2020. [CrossRef]

buildings

MDPI

Article

Effects of Wetting–Drying Cycles on the Macro and Micro Properties of the Cement-Stabilized Soil with Curing Agent

Wenjun Hu [1,2,*], Kun Li [1,2], Wenhao Yin [1,2], Han Zhang [1,2], Yi Xue [1,2], Yutong Han [1,2] and Pingyun Liu [2]

[1] Key Laboratory of Building Structural Retrofitting and Underground Space Engineering (Shandong Jianzhu University), Ministry of Education, Jinan 250101, China; 15216590872@163.com (K.L.); ywh456321@163.com (W.Y.); zhang_h007@163.com (H.Z.); 2023155134@stu.sdjzu.edu.cn (Y.X.); jrcl498498498@163.com (Y.H.)

[2] School of Transportation Engineering, Shandong Jianzhu University, Jinan 250101, China; lpy1802@163.com

* Correspondence: huwenjun@sdjzu.edu.cn

Abstract: Cement-stabilized soil is a commonly used pavement base/bottom base material. Adding a suitable curing agent to cement-stabilized soil can effectively reduce the dosage of cement, meet the strength requirements, and also greatly improve its water stability. In this paper, three kinds of cement dosage (6%, 8%, and 10%) of cement-stabilized soil were selected to add a 0.04% organic liquid curing agent, and then compared with high-dose cement (10% and 12%)-stabilized soil. The influence of wetting–drying cycles on the mechanical properties of the five stabilized soils was discussed. The mineral composition of cement-stabilized soils before and after the addition of a curing agent was analyzed by X-ray diffraction (XRD), and the microscopic morphology of 10% cement-stabilized soils with a curing agent was studied by scanning electron microscopy (SEM). The macroscopic test shows that the unconfined compressive strength of solidified cement-stabilized soil can be divided into three stages with the increase in the times of the wetting–drying cycles, which are the rapid decay stage, stable enhancement stage, and stable decay stage. The wetting–drying stability coefficient first increases, and then decreases with the increase in the times of the wetting–drying cycles. The microscopic test shows that the addition of a curing agent can enhance the content of hydration products in the cement-stabilized soil specimen; at the curing age of 28 d, with the increase in the times of the wet–dry cycles, the structure of the solidified cement-stabilized soil gradually broke down. The surface porosity P and pore diameter d showed an overall upward trend but decreased at the fifth wetting–drying cycle. The pore orientation weakened. The results show that the resistance of cement-stabilized soil with a curing agent is obviously better than that of cement-stabilized soil under wet–dry conditions.

Keywords: curing agent; cement-stabilized soil; wetting–drying cycle; macro and micro properties

1. Introduction

As the base/bottom base material for roads, cement-stabilized soil has the following characteristics compared with cement-stabilized granules [1–6]: (1) more cement; (2) local materials can be used and stone can be saved, which can produce obvious economic and environmental benefits; (3) poor water stability; and (4) low strength. Based on the above points, by adding an organic liquid curing agent to cement-stabilized soil, the goal of reducing the cement dosage, increasing the strength, and improving the water stability can be achieved. The anti-deterioration ability of a material under the action of wetting–drying cycles is an important index to reflect its water stability. When the climate conditions change, the base/bottom base material is subjected to periodic wetting–drying cycles caused by the alternating influence of rainfall and evaporation, and the physical and mechanical properties continue to decay, which will eventually affect the regular use of the road. It is necessary to study the performance changes of solidified cement-stabilized soil in a changing environment (wetting–drying cycle). Liu Kai et al. [7] discussed the change in unconfined compressive strength of cement light soil and geopolymer light soil during wetting–drying cycles, and the research results showed that the wetting–drying cycle had a great influence on geopolymer light soil. Hou ZJ et al. [8] studied the effects of different cement contents on the wetting–drying resistance of Pisha sandstone cement-stabilized soil and found that adding cement can effectively improve the water erosion resistance of the sample, and the optimal cement content is 20%. Li N et al. [9] studied the effect of polypropylene fiber on the wetting–drying degradation resistance of coastal cement-stabilized soil and found that the strength and brittleness index of the sample increased with the increment of the number of wetting–drying cycles. Wang J et al. [10] studied the durability of calcium carbide slag–fly ash solidified shield residuum during wetting–drying cycles and found that the first wetting–drying cycle strongly influenced the compressive strength of the specimen. After the wetting–drying cycles, the stress–strain curve of the specimen changed from strain hardening to strain softening. Yang H et al. [11] analyzed the properties of marine waste silt solidified by PZ-1 under wetting–drying cycles. Neramitkornburi A et al. [12] studied the wetting–drying cycle resistance of the cement solidified clay–fly ash light-weight material and found that the incorporation of fly ash could improve the fluidity and durability of the light-weight material, and the effect was most significant when the fly ash replacement rate was greater than 40%. Kamei T et al. [13] found that adding recycled bassanite and coal ash simultaneously can improve the durability of very soft clay in wet and dry conditions.

In summary, thus far, domestic and foreign scholars have discussed the changes in mechanical properties of different road materials during wet–dry cycles, but relatively few studies have been conducted on the deterioration properties of and microstructure changes in cement-stabilized soil with a curing agent added under wetting–drying cycles. The main test methods used in the microstructure research of solidified soil under wetting–drying cycles are as follows: X-ray diffraction (XRD) and scanning electron microscopy (SEM) [14–16]. Therefore, this study takes cement-stabilized soil and cement-stabilized soil with an organic liquid curing agent as the research objects and evaluates the evolution law and microstructure changes of its anti-deterioration ability during the wetting–drying cycle process through a wet–dry cycle test, unconfined compressive strength test, and microscopic test. The effects of different curing ages and cement contents on the mechanical strength and physical properties of cement-stabilized soil and cement-stabilized soil mixed with a curing agent were discussed. The changes in hydration products in cement-stabilized soil before and after adding a curing agent and the effects of wetting–drying cycles on the microstructure of specimens were studied. The above research results can provide technical support for solidified cement-stabilized soil used as road base/bottom base material.

The highlights are as follows:

- The macro and micro characteristics of solidified cement-stabilized soil under wetting–drying cycles were studied;
- Adding a curing agent can improve the resistance to wetting–drying cycles of cement-stabilized soil;
- The addition of a curing agent can enhance the content of hydration products in cement-stabilized soil.

2. Test Materials and Methods

2.1. Test Materials

(1) Soil sample. The test soil sample was taken from an expressway expansion project in Shandong Province. Figure 1 shows the particle distribution of the soil sample. Table 1 shows the physical property indices of the soil sample.

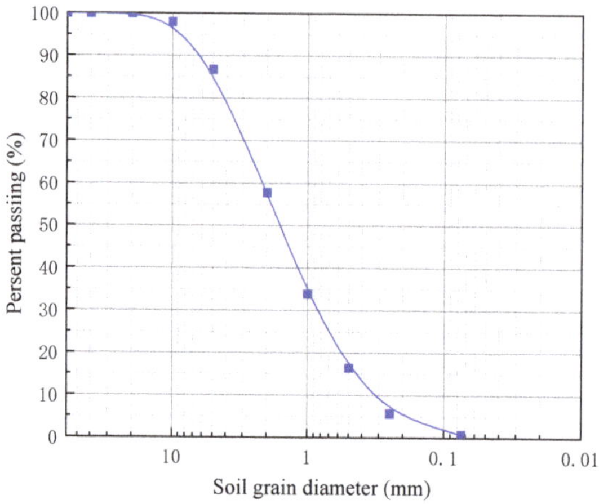

Figure 1. Particle distribution curve of the soil sample.

Table 1. Main physical property indices of the soil sample.

Natural Moisture (%)	PH	Coefficient of Curvature Cc	Coefficient of Nonuniformity Cu	Optimum Moisture (%)	Maximum Dry Density (g/cm³)	Soil Sample (d < 0.5 mm)		
						Liquid Limit (%)	Plastic Limit (%)	Plasticity Index (%)
3.9	8.17	1.0	6.875	12.19	1.9509	42.3	24.6	17.7

The coefficient of nonuniformity C_u is calculated with the following equation:

$$C_u = \frac{d_{60}}{d_{10}} \tag{1}$$

The coefficient of curvature C_c is calculated with the following equation:

$$C_c = \frac{d_{30}^2}{d_{60} \times d_{10}} \tag{2}$$

where d_{60}, d_{30}, and d_{10} = the cumulative percentage of soil weighing less than a certain particle size: 60%, 30%, and 10%, respectively.

(2) Cement. The ordinary silicate P.O42.5 grade cement was used in the test. The main performance indices are shown in Table 2.

Table 2. Main performance indices of cement.

Specific Surface Area (m²/Kg)	Initial Setting Time (min)	Final Setting Time (min)	Stability	3D Bending Strength (MPa)	3D Compressive Strength (MPa)
381	115	184	Eligible	6.2	33.8

(3) Curing agent. The curing agent is an organic liquid, as shown in Figure 2. Diluting the product with a certain proportion of water and applying it to ordinary soil will completely change the hydrophilicity of the soil. The solidified soil not only has good compressive and flexural strength, but also has very prominent water stability.

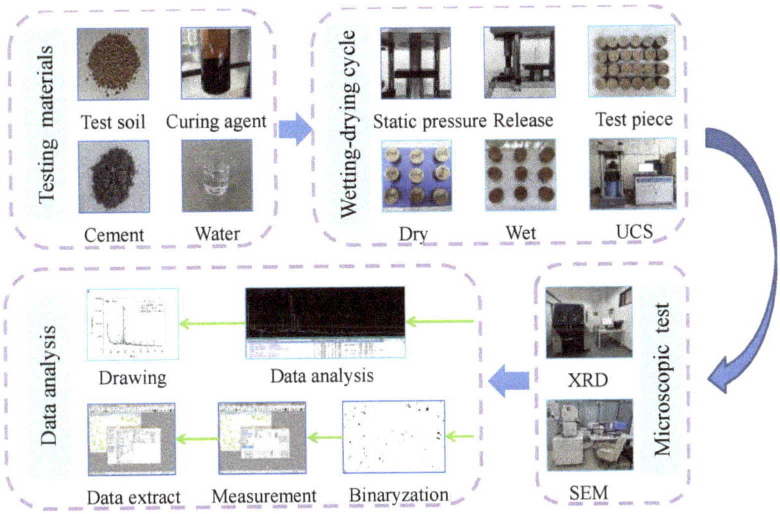

Figure 2. Flow chart of test and data analysis.

2.2. Test Method

Five sets of samples were tested. The sample numbers are shown in Table 3. The specimens used in the wetting–drying cycle test were cylinders (d = 50 mm, and h = 50 mm). The curing age is 7 days and 28 days. After completion of the maintenance, the specimens were placed indoors to air-dry for 24 h, and then wholly soaked 2.5 cm below the water surface for 24 h, which constituted a wetting–drying cycle. Nine wetting–drying cycles were carried out in this experiment. Unconfined compressive strength tests were conducted on the samples after 0, 1, 3, 5, 7, and 9 cycles, respectively. The wetting–drying cycle coefficient (K_w) was used as the evaluation index of the resistance to wetting–drying cycle strength deterioration. The larger the value, the stronger the resistance to deterioration of the specimen. The K_w is calculated with Formula (3):

$$K_w = \frac{R_t}{R_0} \times 100\% \quad (3)$$

where K_w is wetting–drying cycle coefficient (%); R_0 is the unconfined compressive strength (MPa) of the specimen without wetting–drying cycles; and R_t is unconfined compressive strength (MPa) of the specimen at the wetting–drying cycles.

Table 3. Overall test scheme design.

No.	Dosage of Curing Agent (%)	Cement Content (%)	Wetting–Drying Cycles		Microscopic Testing
			Curing Age (d)	Number of Cycles	
S6,0.04	0.04	6	7, 28	0, 1, 3, 5, 7, 9	/
S8,0.04	0.04	8	7, 28	0, 1, 3, 5, 7, 9	/
S10,0.04	0.04	10	7, 28	0, 1, 3, 5, 7, 9	XRD, SEM
S10	0	10	7. 28	0, 1, 3, 5, 7, 9	XRD
S12	0	12	7, 28	0, 1, 3, 5, 7, 9	/

After the wetting–drying cycle test, the small pieces of the samples were selected for the XRD test and scanning electron microscope test. The XRD scanning range 2θ was 5°–80°, and the scanning speed was 2°/min. The data were analyzed using JADE software (MDI Jade 6) [17]. SEM images were first binarized by MATLAB software R2022b [18], and then Image-Pro Plus (IPP) software 6.0 was used to measure and extract the binarized images [19–22]. The overall test scheme design of this study is shown in Table 3, and the test and data analysis process are shown in Figure 2.

3. Mechanical Properties of Stabilized Soil at Different Ages after Wetting–Drying Cycles

3.1. 7 d Curing Age

Figure 3a,b, respectively, show the unconfined compressive strength and wetting–drying cycle coefficient of solidified cement-stabilized soil and cement-stabilized soil under different wetting–drying cycle times at a 7 d curing age.

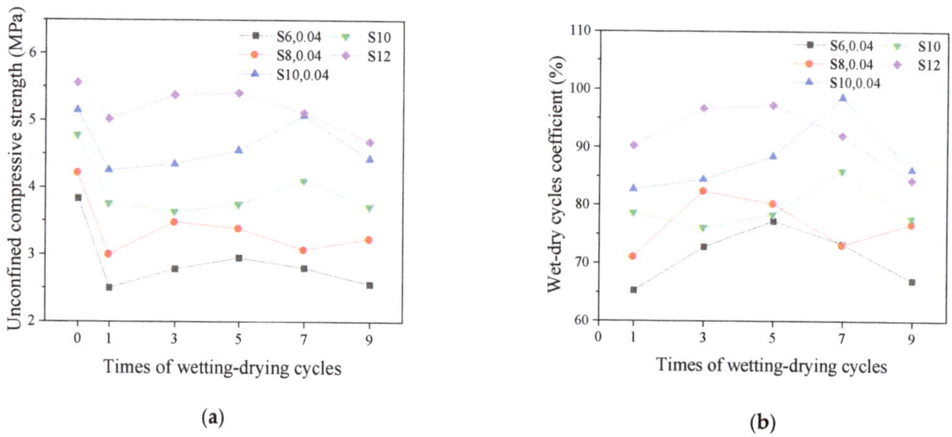

Figure 3. The influence of the number of wetting–drying cycles on unconfined compressive strength and wetting–drying cycle coefficient (7 d): (**a**) unconfined compressive strength; and (**b**) wetting–drying cycle coefficient.

As shown in Figure 3a:

(1) The unconfined compressive strength of the specimen cured for 7 days increases with the increment of cement content under the same wetting–drying cycles. The unconfined compressive strength of S10,0.04 is higher than S10, but lower than S12. It is indicated that adding a 0.04% curing agent is helpful to the resistance of specimens to dry and wet damage.

(2) The development of the unconfined compressive strength of solidified cement-stabilized soil and cement-stabilized soil at the age of 7 days can be divided into the following three stages with the number of wetting–drying cycles:

① The rapid attenuation stage. This mainly occurs during the process of the first wetting–drying cycle. The binding materials in the sample are damaged under the action of the wetting–drying cycles, and the viscosity of the soil mass decreases, which leads to a rapid decrease in the strength of the specimen. The attenuation range of the unconfined compressive strength of the specimen is extensive; Tte maximum is 34.7% (S6,0.04), and the minimum is 10% (S12).

② The stable enhancement stage. This stage mainly occurs during the third, fifth, and seventh wetting–drying cycles. With the increase in the number of wetting–drying cycles, the internal hydration reaction of the specimen was more sufficient, the hydration products were increased, and the pores of the specimen were filled, which led to a gradual increment of the unconfined compressive strength of the specimen. In this stage, the unconfined compressive strength of the S6,0.04, S8,0.04, and S10,0.04 specimens increased by more than 10% compared with the strength after rapid decay. The unconfined compressive strength of the S10 and S12 specimens increased less than 10% compared with the strength after rapid decay.

③ The stable attenuation stage. This stage occurs after the fifth wetting–drying cycle. The internal hydration reaction of the solidified soil sample is completed. With the continuous increment of the number of wetting–drying cycles, the internal structure of the sample is damaged, resulting in the unconfined compressive strength of the solidified cement-stabilized soil decreasing.

As shown in Figure 3b, the wetting–drying cycle coefficients of the specimens cured for 7 days under different dosages increased first, and then decreased with the increment of the number of wetting–drying cycle. At the ninth wetting–drying cycle, the wetting–drying cycle coefficients of the S6,0.04, S8,0.04, S10,0.04, S10, and S12 specimens were 67.02%, 76.78%, 86.21%, 77.82%, and 84.35%, respectively. It shows that the addition of curing agents and the increment of cement content can significantly improve the ability of the specimen to resist failure under a wetting–drying cycle.

Figure 4 shows the appearance changes in the specimens cured for 7 days after nine wetting–drying cycles. As shown in the pictures, the increment of cement content and the addition of curing agents affect the failure state of the sample appearance under wetting–drying cycles. The higher the dosage of the cement and curing agent is, the more intact the specimen's appearance is after 9 wetting–drying cycles. Among them, S6,0.04 experienced the most severe morphological damage, with many surface and edge parts falling off. S12 has a relatively intact shape, with some shallow pits on the surface of the specimen and the edge parts remaining intact.

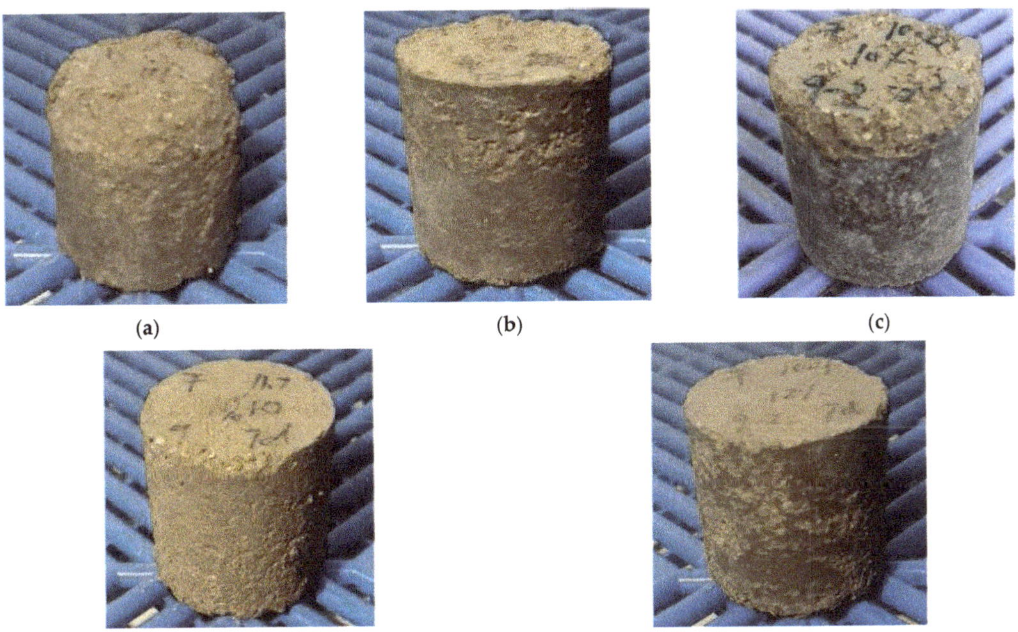

(a) (b) (c) (d) (e)

Figure 4. The appearance changes of the specimens cured for 7 days after 9 wetting–drying cycles: (**a**) S6,0.04; (**b**) S8,0.04; (**c**) S10,0.04; (**d**) S10; and (**e**) S12.

3.2. 28 d Curing Age

Figure 5a,b, respectively, show the unconfined compressive strength and wetting–drying cycle coefficient of the specimens cured for 28 days after nine wetting–drying cycles.

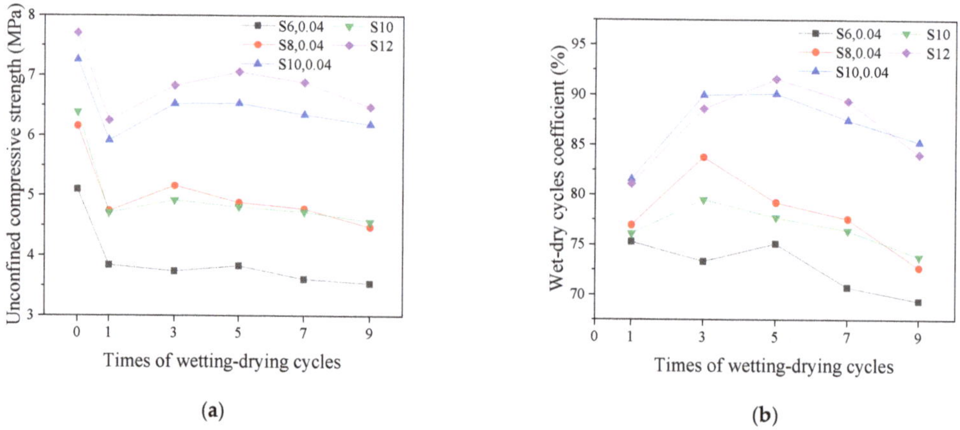

Figure 5. Influence of the number of wetting–drying cycles on unconfined compressive strength and wetting–drying cycle coefficient (28 d): (**a**) unconfined compressive strength; and (**b**) wetting–drying cycle coefficient.

As shown in Figure 5a:

(1) The variation rules of the unconfined compressive strength of the specimens cured for 28 days is the same as those cured for 7 days, which also decrease, then increase, and then decrease. But the duration and the strength growth of the stable strengthening stage decreased. In this stage, the S10,0.04 and S12 specimens showed a relatively noticeable strength growth.

(2) After each wetting–drying cycle, the unconfined compressive strength of S10,0.04 and S12 was almost the same. Under the same number of wetting–drying cycles, the compressive strength curves of the two specimens are also similar, indicating that the addition of a curing agent can reduce the cement content while maintaining the wetting–drying cycle resistance of the sample.

As shown in Figure 5b, the wetting–drying cycle coefficients of the specimens cured for 28 days also showed a tendency to rise up at the beginning and decline at a later stage with the increment of the number of wetting–drying cycles. At the ninth wetting–drying cycle, the wetting–drying cycle coefficients of S6,0.04, S8,0.04, and S10,0.04 were 69.4%, 72.7%, and 85.3%, respectively. It can be seen that the higher the cement dosage, the stronger the specimen's resistance to the wetting–drying cycle damage. S10,0.04 and S12 had similar wetting–drying cycle coefficients under different wetting–drying cycles, and the change trend was the same. At the ninth wetting–drying cycle, the wetting–drying cycle coefficient of sample S10,0.04 was slightly higher, indicating that, with the joining of a curing agent, the ability of the sample to resist wetting–drying cycle erosion was also enhanced.

The influence of the number of wetting–drying cycles on the appearance change in specimen S10,0.04 cured for 28 d is shown in Figure 6. As can be seen from Figure 6, the surface of S10,0.04 did not change significantly after the first wetting–drying cycle, and the edge fell off slightly after the third wetting–drying cycle. The damage was the most serious after the ninth wetting–drying cycle, with shallow pits on the surface and edge caused by repeated erosion. Due to the water loss and absorption at the edge of the sample surface, the particle shedding first occurred in this area [23].

Figure 6. The influence of the number of wetting–drying cycles on the appearance changes of S10,0.04 (28 d): (**a**) one wet–dry cycle; (**b**) three wetting–drying cycles; (**c**) five wetting–drying cycles; (**d**) seven wetting–drying cycles; and (**e**) nine wetting–drying cycles.

4. Mechanical Properties of Stabilized Soil at Different Ages after Wetting–Drying Cycles

4.1. XRD Analysis

An XRD phase analysis was conducted on S10 and S10,0.04 cured for 28 days. Figure 7 shows that the mineral composition of the two soil samples mainly include quartz, calcium silicate hydrate, ettringite, etc. Both S10 and S10,0.04 have a crystal diffraction peak. However, the diffraction peaks of AFt and C-S-H crystals in S10,0.04 increase. It can be seen that the main change in the sample before and after the joining of the curing agent is the content of hydration products. The curing agent plays a role in accelerating the cement hydraulic and pozzolanic reaction.

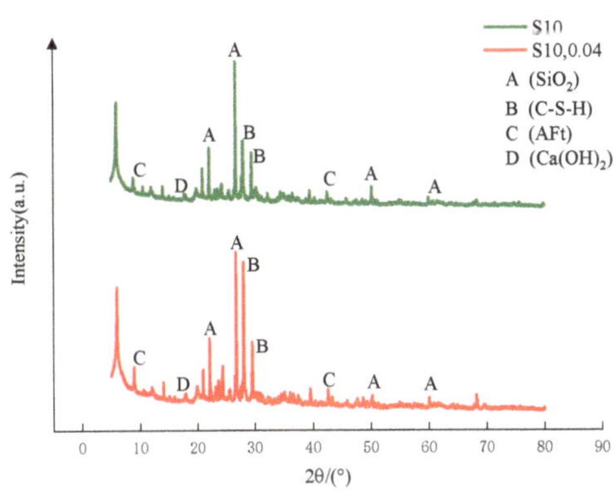

Figure 7. XRD phase analysis of S10 and S10,0.04 (28 d).

4.2. SEM Analysis

S10,0.04 cured for 28 d was tested by SEM after zero, one, five, and nine wetting–drying cycles, at magnifications of 200 times and 2000 times. The SEM results are shown in Figures 8 and 9. As can be seen from the figures, with the increment of the wetting–drying cycles, soil particles were broken, the pore structure was gradually loosened, and the cementation between soil grain was obviously weakened. The contact mode of particles changed from close contact to point or surface contact, and the structure became loose. The number of large pores in the soil pores obviously increased, and the strength property of the soil was weakened. This indicates that S10,0.04 experienced repeated expansion and contraction deformation during the wetting–drying cycles, soil particles were scoured by water flow, and the corresponding material components and hydration products in the soil continued to migrate and disperse, resulting in changes in soil micromorphology and severe damage to mechanical properties.

(**a**)

(**b**)

(**c**)

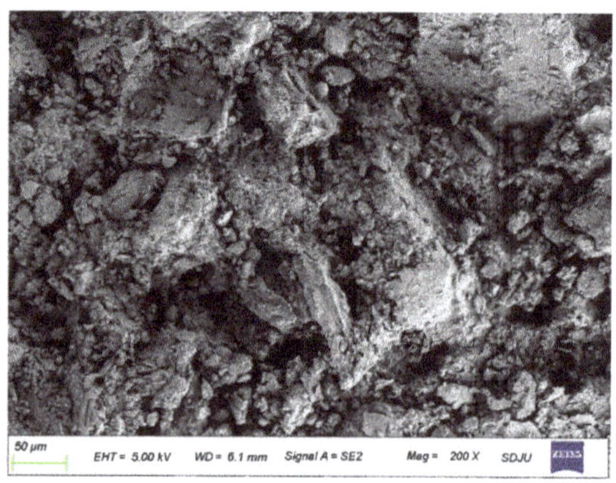

(**d**)

Figure 8. S10,0.04. SEM image ×200 times after wetting–drying cycles: (**a**) zero wetting–drying cycle; (**b**) one wetting–drying cycle; (**c**) five wetting–drying cycles; and (**d**) nine wetting–drying cycles.

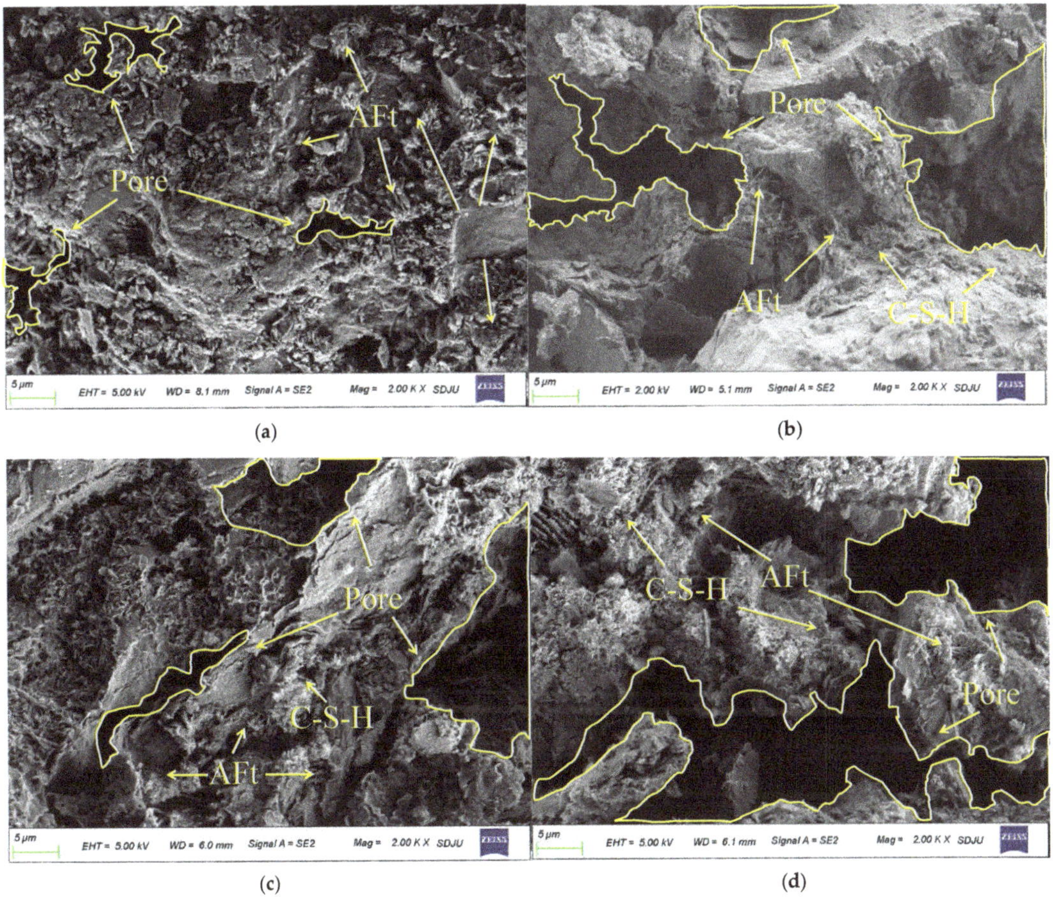

Figure 9. S10,0.04. SEM image ×2000 times after wetting–drying cycles: (**a**) zero wetting–drying cycle; (**b**) one wetting–drying cycle; (**c**) five wetting–drying cycles; and (**d**) nine wetting–drying cycles.

4.2.1. Surface Porosity P

Surface porosity is calculated according to Formula (4):

$$P = \frac{\sum_{i=1}^{N} S_i}{S_{total}} \tag{4}$$

where N is the total number of pores in the electron microscope picture; S_i is the area of the ith pore (μm^2); and S_{total} is the area of the picture (μm^2).

The variation of the surface porosity P of S10,0.04 under different wetting–drying cycles is shown in Figure 10. The surface porosity of S10,0.04 showed an overall upward trend with the increment of the number of wetting–drying cycles, but the surface porosity decreased after five wetting–drying cycles. This indicates that, under wetting–drying cycles, the internal structure of the sample gradually broke down, and pores and cracks increased continuously. After the fifth wetting–drying cycle, the decrease in surface porosity was due to the wetting–drying cycle process promoting the internal hydration reaction, and the hydration products generated filled the expanded pores and cracks. After the fifth wetting–drying cycle, the hydration reaction was nearly complete. AFt, C-H-S, and other hydration products no longer increased significantly, and the surface porosity of S10,0.04 increased again.

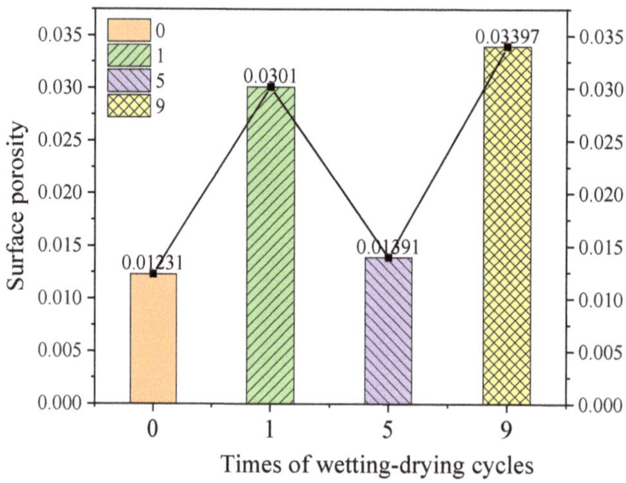

Figure 10. Changes in surface porosity of S10,0.04 under different wetting–drying cycles.

4.2.2. Pore Diameter d

The pore diameter distribution of S10,0.04 under different wetting–drying cycles is shown in Figure 11a, and the change in the average pore diameter is shown in Figure 11b. The pore diameter of S10,0.04 was mainly concentrated within 1~5 um under the wetting–drying cycles. With the increment of the number of wetting–drying cycles, the tiny pores with a diameter of less than 1 um gradually decreased; in contrast, the large pores with a diameter of more than 10 um continuously increased, and the average pore diameter showed an upward trend. After the ninth dry–wet cycle, the average pore size of S10,0.04 is 1.41 times that of the sample without wetting–drying cycles, and 1.14 times that of the sample after the first wetting–drying cycle.

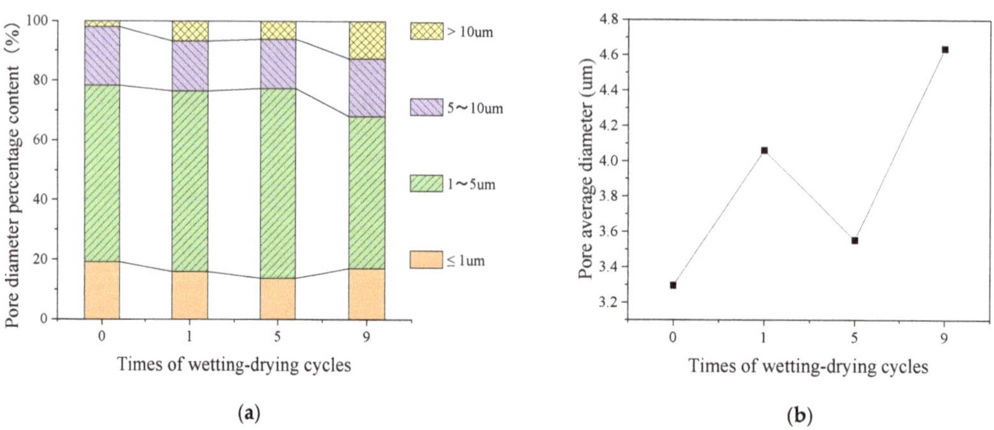

Figure 11. Changes in pore diameter of S10,0.04 under different wetting–drying cycles: (**a**) distribution pattern of pore diameter; and (**b**) variation pattern of average pore diameter.

According to the pore diameter distribution of S10,0.04 under different wetting–drying cycles, the relationship between pore diameter distribution and surface porosity was established, and the fitted results are demonstrated in Figure 12. As can be seen from Figure 12, ≤1 um and 1~5 um apertures have the worst correlation with surface porosity. The contribution rate of ≤1 um and 1~5 um apertures to the porosity increase is negative.

>10 um and 5~10 um apertures have a higher correlation with surface porosity. The contribution rate of >10 um aperture to the increment of porosity is positive, and the contribution rate of the 5~10 um aperture is negative. This shows that the rise in surface porosity is mainly related to the >10 um aperture.

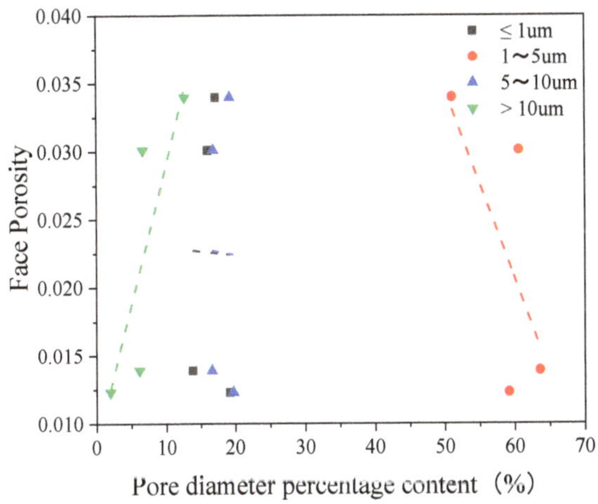

Figure 12. Relationship between pore diameter distribution and surface porosity under different wetting–drying cycles.

According to the pore diameter distribution of S10,0.04 under different wetting–drying cycles, the relationship between pore diameter distribution and unconfined compressive strength was established, and the fitted results are demonstrated in Figure 13. It can be seen from Figure 13 that the four aperture ranges all had an impact on the unconfined compressive strength. The pore diameter ≤1 um, 1 um~5 um, and 5 um~10 um have a low correlation with the strength, and the correlation index R^2 is 0.27064, 0.03754, and 0.31628, respectively, and they are all positively correlated. The pore diameter >10 um has a high correlation with the strength, and the correlation index R^2 is 0.50022, showing a negative correlation.

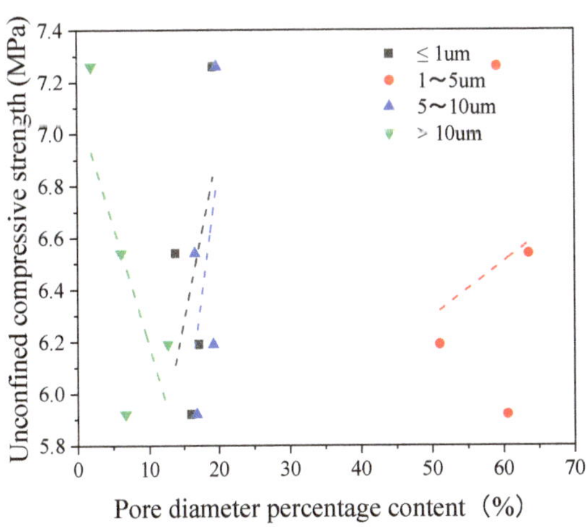

Figure 13. The relationship between pore diameter distribution and unconfined compressive strength under different wetting–drying cycles.

4.2.3. Pore Abundance

The porosity is calculated according to Formula (5):

$$C = \frac{B}{L} \quad (5)$$

where B is the pore short-axis length (um); L is the long-axis length of the pore (um).

The pore abundance distribution of S10,0.04 under different wetting–drying cycles is shown in Figure 14a, and the change in average pore abundance is shown in Figure 14b. As can be seen from Figure 14a, the porosity of S10,0.04 was mainly concentrated in the range of 0.4~0.6, and the porosity of 0.0~0.2 accounts for a very tiny proportion during the wetting–drying cycles, indicating that most of the pore shapes tended to be elliptical and elongated pores were very rare. As shown from Figure 14b, the average pore abundance of S10,0.04 fluctuated within the small range of 0.4~0.6 under different wetting–drying cycles, indicating that the wetting–drying cycles had little influence on the pore shape.

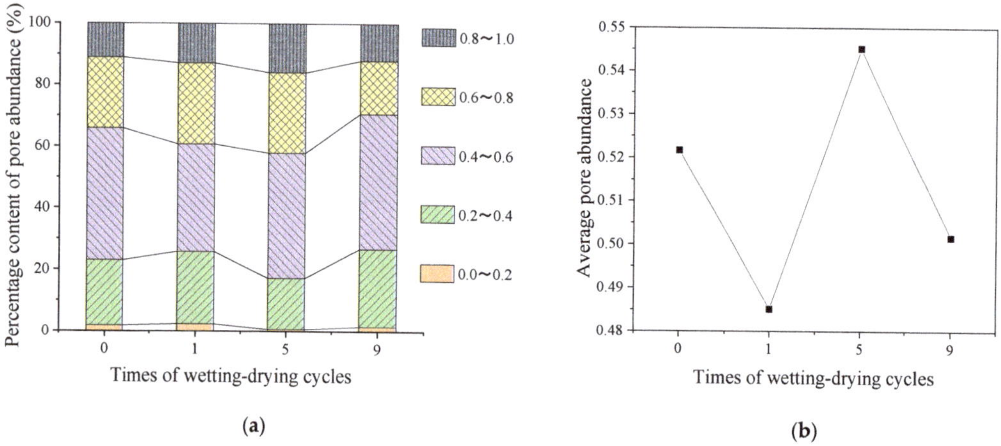

Figure 14. Changes in pore abundance of S10,0.04 under different wetting–drying cycles: (**a**) distribution rule of pore abundance; and (**b**) variation rule of average pore abundance.

4.2.4. Fractal Dimension

The fractal dimension is calculated according to Formula (6), which reflects the complicacy of the pore shape. Its value is generally between 1 and 2. The greater the value, the more multiple the pore profile [24,25].

$$D = -\lim_{\varepsilon \to 0} \frac{\ln N(\varepsilon)}{\ln \varepsilon} \quad (6)$$

where ε is the pore diameter (μm); and $N(\varepsilon)$ is the percentage (%) of the total number of pores greater than this diameter.

The variation law of the mean fractal dimension of S10,0.04's pores under different wetting–drying cycles is shown in Figure 15. The mean fractal dimension of S10,0.04's pores decreased first, then increased with the increment of the number of wetting–drying cycles, mainly between 1.108 and 1.115. The average fractal dimension of pores after the ninth wetting–drying cycle increased by 0.36% compared with that without wetting–drying cycles, and by 0.51% compared with that after the first wetting–drying cycle, indicating that the wetting–drying cycle process had a limited influence on the average fractal dimension of pores.

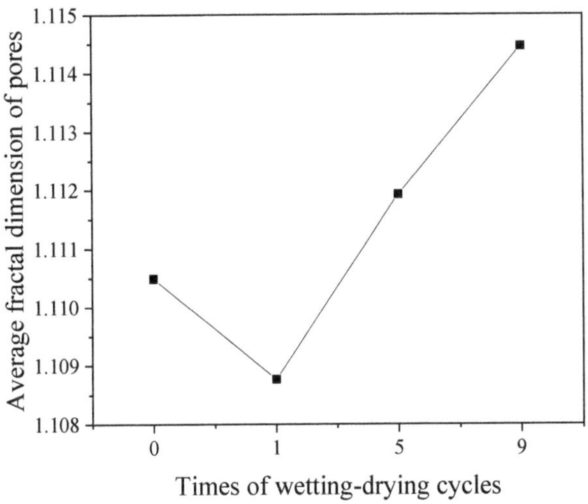

Figure 15. Changes in S10,0.04's mean fractal dimension of pores under different wetting–drying cycles.

4.2.5. Pore Orientation Angle

The angle between the long axis of the pores in the binarized picture and the Y-axis of the picture co-ordinate system is called the pore orientation angle [26]. The distribution frequency of the pores in a specific direction is equally divided in the interval range of 0~180° by every 10°, so that, within the range of 0~180°, the orientation frequency of the pores in the i interval can be calculated according to Formula (7) [27]:

$$F_i = \frac{m_i}{M} \times 100\% \quad (7)$$

where m_i is the number of pores in the i interval in the binary picture; and M is the total number of pores.

In the range of 0°~360°, the data are processed in mirror symmetry. The pore orientation frequency of S10,0.04 under different wetting–drying cycles is shown in Figure 16. As seen from Figure 16, the porosity has a specific orientation, and its orientation angle is mainly distributed in the range of 90°~100°. With the increment of the number of wetting–drying cycles, the pore orientation gradually decreased, and the pore distribution became more disordered, showing a trend of uniform distribution at all angles. However, the proportion of the orientation angle within 90°~100° was still considerable.

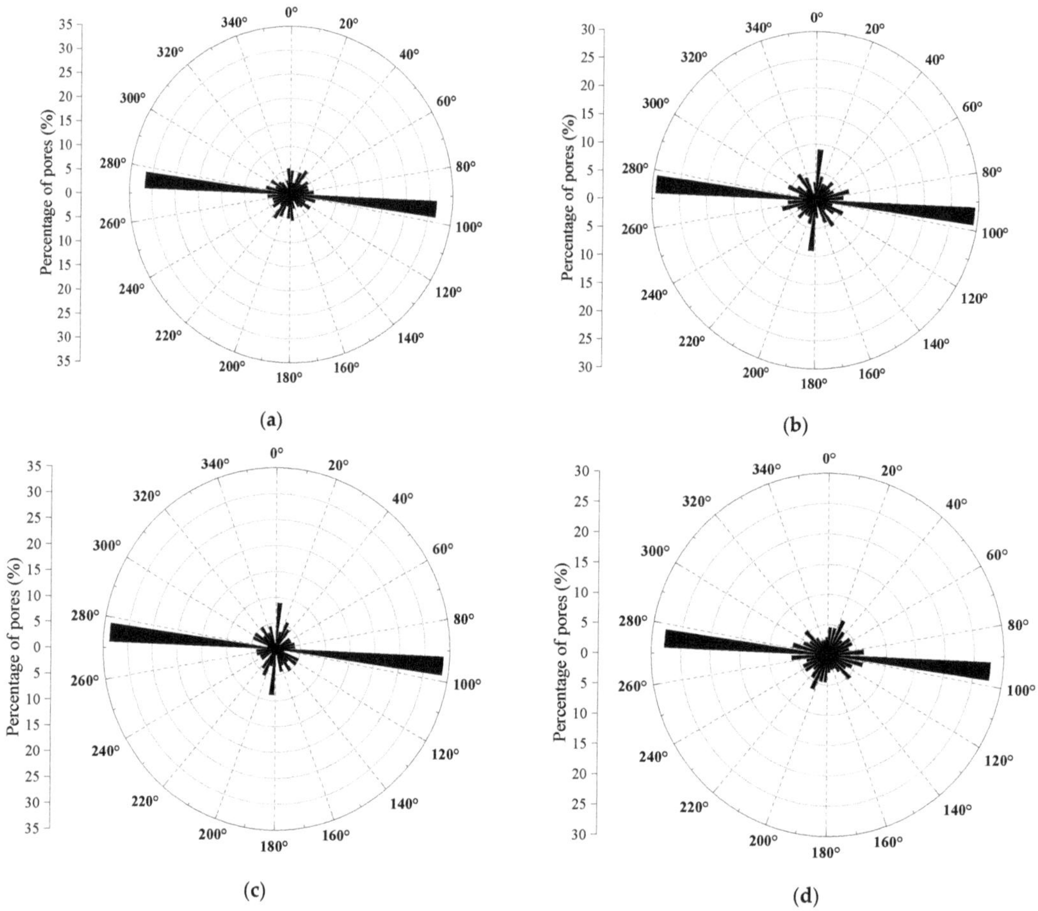

Figure 16. Pore orientation frequency of S10,0.04 under different wetting–drying cycles: (**a**) pore orientation frequency after zero wetting–drying cycle; (**b**) pore orientation frequency after one wetting–drying cycle; (**c**) pore orientation frequency after five wetting–drying cycles; and (**d**) pore orientation frequency after nine wetting–drying cycles.

5. Conclusions

The macro and micro characteristics of solidified cement-stabilized soil under wetting–drying cycles were studied. The effects of wetting–drying cycles on the mechanical properties of solidified cement-stabilized soil (S6,0.04, S8,0.04, and S10,0.04) and cement-stabilized soil (samples S10 and S12) were investigated through compressive strength tests. The mineral composition changes of cement-stabilized soil before and after the addition of the curing agent (S10, and S10,0.04) were analyzed by X-ray diffraction (XRD), and the microscopic morphological changes of solidified cement-stabilized soil (S10,0.04) under different wetting–drying cycles were further studied by scanning electron microscopy (SEM). The conclusions are as follows:

(1) With the increase in the number of wetting–drying cycles, the unconfined compressive strength of all samples decreased first, and then increased, and decreased finally. It indicates that the specimen underwent significant damage in the early stage of the wetting–drying cycles, the hydration reaction was enhanced in the middle period in the immersed water environment, and the strength could be restored. In the later stage, the

hydration reaction was completed, and no hydration products were produced. Therefore, the unconfined compressive strength of the specimen declined again.

(2) With the increase in wetting–drying cycles, the wetting–drying cycle coefficient of solidified cement-stabilized soil first increased, and then decreased, the specimen structure was gradually damaged, the surface shallow pits gradually increased, and the edge parts continued to fall off. The erosion of S6,0.04 was the most serious, and the increment of cement content and the incorporation of the curing agent could effectively improve its resistance to wetting–drying erosion.

(3) The addition of the curing agent effectively promoted the hydration reaction and improved the resistance to deterioration of the samples during the wetting–drying cycles. The unconfined compressive strength, wetting–drying cycle stability coefficient, and mass change rate of S10,0.04 were superior to that of S10. The performance of S10,0.04 was very similar to that of S12.

(4) With the increase in the times of the wetting–drying cycles, the surface porosity P and pore diameter d of S10,0.04 showed an overall upward trend, and decreased at the fifth wetting–drying cycle. The pore orientation was weakened, and the changes in pore abundance and fractal dimension were small. The unconfined compressive strength of S10,0.04 was inversely proportional to the pore percentage content of a >10 um pore diameter, and proportional to the pore percentage contents of a $\leq$1 um, 1 um~5 um, and 5 um~10 um pore diameter.

Author Contributions: Conceptualization, and funding acquisition, W.H.; writing—original draft preparation, and conceptualization, K.L.; writing—review and editing, and validation, W.Y.; formal analysis, H.Z.; investigation, Y.X. and Y.H.; data curation, P.L. All authors have read and agreed to the published version of the manuscript.

Funding: This research was supported by the Natural Science Foundation of Shandong Province (Grant No. ZR2021ME238) and the Project of Shandong Province Higher Educational Science and Technology Program (Grant No. J18KA216).

Data Availability Statement: Data is contained within the article.

Conflicts of Interest: The authors declare no conflicts of interest.

Nomenclature

XRD	X-ray Diffraction
SEM	Scanning Election Microscopy
UCS	Unconfined Compressive Strength
d	day

References

1. Buritatum, A.; Horpibulsuk, S.; Udomchai, A.; Suddeepong, A.; Takaikaew, T.; Vichitcholchai, N.; Horpibulsuk, J.; Arulrajah, A. Durability Improvement of Cement Stabilized Pavement Base Using Natural Rubber Latex. *Transp. Geotech.* **2021**, *28*, 100518. [CrossRef]
2. Jiang, N.; Wang, C.M.; Wang, Z.P.; Li, B.; Liu, Y.A. Strength Characteristics and Microstructure of Cement Stabilized Soft Soil Admixed with Silica Fume. *Materials* **2021**, *14*, 1929. [CrossRef] [PubMed]
3. Zhou, M.K.; Liu, X.Y.; Chen, X.; Gao, P. Study on Strength, Water Stability, Shrinkage, and Microstructure of CFB Slag Modified Cement Stabilized Clay. *Materials* **2021**, *14*, 7460. [CrossRef] [PubMed]
4. Ji, J.; Liang, B.; Han, B.Y.; Zhi, S.; Wang, J.N.; Yu, H.C. Review on soil solidified technologies in road engineering in China. *J. Traffic Transp. Eng.* **2023**, *23*, 47–66. (In Chinese)
5. Hu, Z.H.; Wang, Y.; Wang, C.J. Study on mechanical properties of dredged silt stabilized with cement and reinforced with alginate fibers. *Case Stud. Constr. Mater.* **2023**, *18*, e01977. [CrossRef]
6. Zhang, Z.C.; Omine, K.; Flemmy, S.O. Evaluation of the improvement effect of cement-stabilized clays with different solidifying agent addition and water content. *J. Mater. Cycles Waste Manag.* **2022**, *14*, 2291–2302. [CrossRef]
7. Liu, K.; Li, R.M.; Du, Y.J.; Wei, M.L. A durability experimental study of light-weight soil subjected to wetting-drying cycles and sodium sulfate soaking. *Rock Soil Mech.* **2015**, *36* (Suppl. S1), 362–366.

8. Hou, Z.J.; Wang, A.T.; Yao, Z.Y. Study on the mechanical property and wetting-drying cycle durability of cement foamed light-weight soil. *J. Xi'an Univ. Archit. Technol.* **2021**, *53*, 80–85.
9. Li, N.; Zhao, W.Q.; Gu, J.H.; Zhao, S.W.; Wang, W.; Jiang, P. Study on mechanical properties of coastal cement soil modified by fiber under dry-wet cycle. *Yangtze River* **2022**, *53*, 169–174.
10. Wang, J.; Fan, Y.; Xiong, X.; Zhao, F. Stabilization of Shield Muck Treated with Calcium Carbide Slag–Fly Ash. *Buildings* **2023**, *13*, 1707. [CrossRef]
11. Yang, H.; Zhu, J.F.; Tao, Y.L.; Wang, Z.; Zheng, Q. Effect of the Dry-Wet Cycle on the Performance of Marine Waste Silt Solidified by Calcium Carbide Residue and Plant Ash. *J. Mar. Sci. Eng.* **2022**, *10*, 1442. [CrossRef]
12. Neramitkornburi, A.; Horpibulsuk, S.; Shen, S.L.; Chinkulkijniwat, A.; Arulrajah, A.; Disfani, M.M. Durability against wetting–drying cycles of sustainable Light-weight Cellular Cemented construction material comprising clay and fly ash wastes. *Constr. Build. Mater.* **2015**, *77*, 41–49. [CrossRef]
13. Kamei, T.; Ahmed, A.; Shibi, T. The use of recycled bassanite and coal ash to enhance the strength of very soft clay in dry and wet environmental conditions. *Constr. Build. Mater.* **2013**, *38*, 224–235. [CrossRef]
14. Wang, Z.; Wei, B.P.; Wu, X.H.; Zhu, H.; Wang, Q.; Xiong, Z.; Ding, Z. Effects of dry-wet cycles on mechanical and leaching characteristics of magnesium phosphate cement-solidified Zn-contaminated soils. *Environ. Sci. Pollut. Res.* **2021**, *28*, 18111–18119. [CrossRef] [PubMed]
15. Liang, S.H.; Wang, Y.X.; Feng, D.L. Experimental Study on Strength and Dry-Wet Cycle Characteristics of South China Coastal Soft Soil Solidified by Cement Collaborating Sand Particles. *Appl. Sci.* **2023**, *13*, 8844. [CrossRef]
16. Liu, Y.; Lu, H.J.; Liu, M.Y.; Cai, L.; Wei, N.; Liu, Y. Microanalytical characterizations, mechanical strength and water resistance performance of solidified dredged sludge with industrial solid waste and architecture residue soil. *Case Stud. Constr. Mater.* **2022**, *17*, e01492. [CrossRef]
17. Zhang, S.; Liu, Q.F.; Zhang, H.; Ma, R.; Li, K.; Wu, Y.; Teppen, B.J. Structural order evaluation and structural evolution of coal derived natural graphite during graphitization. *Carbon* **2020**, *157*, 714–723. [CrossRef]
18. Wang, H.; Cao, Y.K.; Ren, J.X. Influence of Dry-wet Cycle on Crack and Strength of Silty Sand Improved Expansive Soil. *J. Archit. Civ. Eng.* **2022**, *39*, 213–221.
19. Huang, Y.H.; Chen, Y.; Zhu, X.; Wu, Z.Q.; Zhu, R.; Wang, S.; Wu, M. Experimental study and micro-mechanism analysis of freeze-thaw performance of expansive soils improved by phase-change materials. *Chin. J. Geotech. Eng.* **2021**, *43*, 1994–2002.
20. Li, Y.T.; Yuan, L. Research on Micro-Mechanism of Cement Modified Saline Soil Strength Development. *Highway* **2013**, *3*, 164–168.
21. Peng, Y.S.; Ou, X.D.; Ji, F.L. Physical and Mechanical Properties and Meso-characteristics of Foamed Mixture Lightweight Soil with Bauxite Tailings. *Mater. Rep.* **2022**, *36*, 128–133.
22. Li, G.W.; Wang, J.Y.; Chen, W.; Wu, J.T.; Cao, X.S.; Wu, S.F. Influences of wetting-drying cycles on expansive soils improved with disintegrated sandstone with different particle size groups. *Chin. J. Geotech. Eng.* **2022**, *44*, 643–651.
23. He, Z.J.; Chai, S.X.; Li, M. Effect of wetting and drying cycles on the compressive strength of oil-contaminated soil treated with lime and fly ash. *J. Eng. Geol.* **2018**, *26*, 438–444.
24. Yang, H.; Zhu, J.F.; Tao, Y.L. Research on the micro-characteristics of solidified waste sludge under the effect of three factors. *Hydrogeol. Eng. Geol.* **2022**, *49*, 91–99.
25. Li, J.D.; Wang, X.; Zhang, Y.J.; Jiang, D.J.; Liu, D.; Wang, J.; Steven. A Mechanism Study of Trial Loess Reinforced by F1 Ionic Soil Stabilizer on Curing Mechanism and Strength Characteristics. *Mater. Rep.* **2021**, *35*, 6100–6106.
26. Cai, S.; Ruan, Y.F.; Li, P.F.; Zhu, Q.; Yan, M. Shear characteristics and microstructural changes of peat soil. *China Earthq. Eng. J.* **2022**, *44*, 1366–1374.
27. Liu, K.; Ye, W.J.; Gao, H.J.; Dong, Q. Multi-scale effects of mechanical property degradation of expansive soils under drying-wetting environments. *Chin. J. Rock Mech. Eng.* **2020**, *39*, 2148–2159.

buildings

MDPI

Article

Evaluation of Interlayer Reinforcement Effectiveness in Road Pavement Rehabilitation Using FEM Modeling and Fracture Mechanics Analysis

Arianna Antoniazzi , Gianluca Ravizzoni, Cecilia Schiavone, Maurizio Crispino and Emanuele Toraldo *

Department of Civil and Environmental Engineering, Politecnico di Milano, 20133 Milano, Italy; arianna.antoniazzi@polimi.it (A.A.); gianluca.ravizzoni@mail.polimi.it (G.R.); cecilia.schiavone@mail.polimi.it (C.S.); maurizio.crispino@polimi.it (M.C.)
* Correspondence: emanuele.toraldo@polimi.it

Abstract: In this paper, the effectiveness of reinforcements for flexible pavements is evaluated through an analysis of reflective cracking. Different stiffness and thickness reinforcements are considered for the rehabilitation of an already cracked pavement. The effect of the reinforcement is assessed from two different perspectives: (i) the ability to reduce stresses in the rehabilitated pavement layers, and (ii) the capacity to mitigate the crack propagation from deeper layers. A finite element model (FEM) is adopted to study the stress and strain state of the pavement layers. The pavement model has been properly validated, transitioning from a simply supported beam scheme to an elastic multilayer model. In addition, to represent crack propagation, fracture evolution is analyzed using Linear Elastic Fracture Mechanics (LEFMs) and Paris' law. The effect of different reinforcements on the pavement is then simulated. The results show that the reinforcement performance is strictly dependent on the interlayer thickness and stiffness. In particular, high stiffness reinforcements (geomembranes) show increasing effectiveness with stiffness, both in terms of reflective cracking and stress reduction. Conversely, low stiffness reinforcements (SAMIs) show a variable trend with the stiffness modulus. In fact, extremely low stiffness is effective in slowing down crack propagation but is detrimental to the wearing course's stress condition. However, as the stiffness increases, the likelihood of cracking in the wearing course decreases, though only a small beneficial effect is registered for crack propagation in the base layer.

Keywords: pavement cracking; pavement reinforcement; SAMI; geomembrane; fracture mechanics; finite element modeling; elastic multilayer model

1. Introduction

Fatigue cracking is one of the most common distress mechanisms in asphalt pavements. It consists of the initiation and propagation of cracks due to the repeated loading of the pavement [1]. The fracture severity and speed of propagation depend on various intrinsic and extrinsic parameters of hot mix asphalt pavement, such as layer thickness, mechanical properties, applied load, and temperature [2]. Cracked pavement causes the loss of regularity, friction, and comfort for vehicles, with significant safety implications.

One possible intervention to slow down reflective cracking is the implementation of reinforcement layers. To place the reinforcement layer on an initially cracked pavement, it is necessary to partially mill the pavement so that the final level of the pavement is not changed. An interlayer of variable thickness is then placed, and a final asphalt overlay is applied to restore the pavement surface characteristics. The effectiveness of the reinforcement depends on the cracking extension in the original pavement, the thickness and stiffness of the reinforcement, the adhesion of the reinforcement to the adjacent layers, and the accuracy of the execution. This paper focuses mainly on the interlayer reinforcement properties to define the optimal solution in terms of reinforcement thickness and stiffness.

Until now, several reinforcement materials have been considered for applications, but a generic approach aimed at comparing different alternatives has not yet been developed. In fact, the objective of this paper is to define a range of possible effective reinforcement solutions, in terms of thickness and stiffness, that could be obtained by adopting various materials available on the market.

2. Literature Review

The fatigue resistance of pavements depends on a combination of factors; therefore, machine learning and predictive methods have been used to describe it [3].

Various techniques can be adopted to prevent crack initiation [4] or to delay crack propagation [5]. Some methods even involve the adoption of recycled materials [6].

In particular, to delay fatigue cracking propagation, different reinforcement interventions could be implemented. The process mainly consists of milling part of the surface course to place a reinforcement interlayer, over which an overlay is applied. A widely used method consists of placing a geocomposite interlayer between the existing pavement and a new asphalt overlay. The effectiveness of the geocomposite in delaying crack initiation and propagation has been evaluated through laboratory tests, showing that material properties play a crucial role [7]. For instance, the suitability of the mechanical properties of reinforcing materials could be assessed for both natural and synthetic textiles [8]. It should be noted that the effect of geosynthetics in delaying crack initiation and propagation is usually evaluated by laboratory tests, which do not take into account a previously cracked pavement [9]. Moreover, laboratory tests conducted to characterize fatigue resistance are affected by great dispersion and require a large sample size [10].

The Stress Absorbing Membrane Interlayer (SAMI) could also be implemented as interlayers. Various studies in the literature have analyzed the effectiveness of SAMIs through laboratory tests designed to simulate fatigue cracking through accelerated loading tests [11]. The effectiveness of SAMIs might depend on various boundary conditions, and it was shown that their introduction is not always beneficial to the road pavement. In particular, the benefits of the SAMI depend on its thickness and the shear stiffness of the interlayer, especially when applied to a cracked pavement [12,13]. Moreover, depending on the temperature, the effect of the SAMI could vary, and when dealing with visco-elastic materials, such as asphalt pavements, a change in temperature results in a change in stiffness. This confirms the importance of considering both reinforcement thickness and stiffness as study variables.

SAMIs are low stiffness interlayers that are able to absorb stress due to their deformability. A Finite Element Analysis, which was used to evaluate SAMIs, revealed that low stiffness is necessary, but extremely low stiffness could lead to undesirable effects, such as excessive strain in the overlay [14].

In addition, SAMIs that employ innovative materials, such as a fiber reinforced SAMI [15] and asphalt rubber SAMI [16], have also been considered.

Bonding efficiency is another important aspect of interlayer reinforcement. In fact, a defect in interlayer bonding could compromise the effectiveness of the reinforcement [17,18]. Researchers investigated the feasibility of applying geocomposites to a milled surface and showed that they can be successfully applied if properly implemented [19]. It was demonstrated that bonding properties depend on the reinforcement system type [20]. Moreover, the adoption of innovative materials (such as a basalt fiber mesh geotextile [21]) to improve adhesion has been taken into account.

Finite Elements Models (FEMs) have been adopted in the literature to describe crack propagation by means of Paris' law [22]. This method involves an energetic approach to describe fatigue cracking [23] based on the introduction of a Stress Intensity Factor (SIF) and the J-integral [24].

3. Research Objectives

In this study, the effectiveness of the reinforcement is evaluated according to the state of the art. In particular, as interlayer thickness and stiffness are essential characteristics, different values for these variables have been taken into account. To simulate the effect on crack propagation and stress condition in the pavement layers, a FEM method is adopted. The simulations consider an already cracked pavement as the initial condition, representing a more realistic situation for reinforcement implementation. In fact, crack initiation occurs at a localized defect and the number of load cycles inducing initiation is strictly dependent on the specific defect and is thus influenced by randomness [25]. However, the crack propagation trend depends on the overall pavement layer properties. Therefore, the focus of this study is to describe and simulate the crack propagation phase. In detail, crack propagation is simulated in accordance with Paris' law, which, based on the literature, is the most widely adopted approach to date.

In addition, the unreinforced pavement condition is also considered for crack propagation simulations. This allows the performance of the specific reinforcement type to be evaluated by comparing it with the zero-intervention solution. As a result, a set of appropriate reinforcement characteristics could be determined.

Specifically, the effectiveness of each type of reinforcement is assessed on the basis of the following criteria: (i) the ability to limit the stress condition in the asphalt overlay and prevent it from cracking, and (ii) the ability to slow down and delay the propagation of cracks in the initially cracked pavement. In fact, the cracking of the overlay would again compromise the surface properties of the pavement in terms of friction and therefore safety.

4. Materials and Methods

The purpose of this study is to assess the effectiveness of asphalt pavement reinforcement in counteracting reflective cracking. It is assumed that the rehabilitation intervention is carried out on an initially cracked pavement to limit the negative impact on user safety.

4.1. Pavement Rehabilitation and Reinforcement Layer

Road pavement rehabilitation involves the partial milling of the cracked pavement layers, the application of a reinforcement interlayer, and a final asphalt overlay to restore the surface characteristics. In particular, various reinforcement interlayer typologies have been taken into account in terms of stiffness and thickness. To evaluate the effectiveness of different types of reinforcement, it is necessary to simulate crack propagation scenarios in pavements without and with reinforcement. By comparing the results for the different reinforcements, the most effective solutions in terms of thickness and stiffness can be identified. In particular, the pavement layouts without reinforcement and with reinforcement are shown in Figure 1a and Figure 1b, respectively.

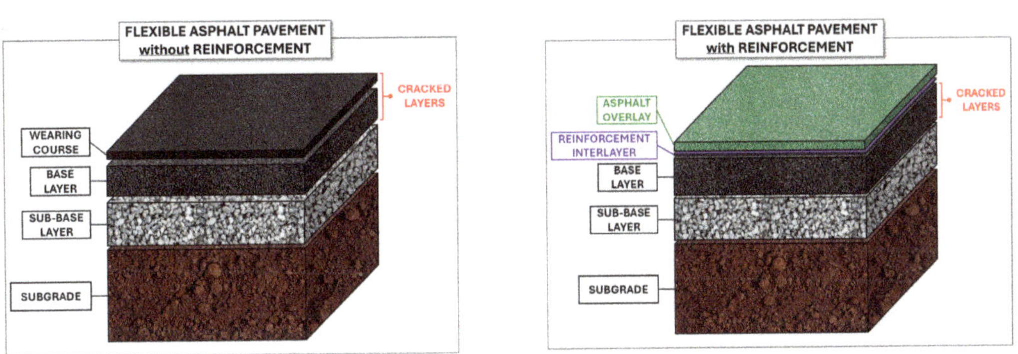

(a) (b)

Figure 1. Analyzed pavement scheme: (**a**) without reinforcement; (**b**) with reinforcement.

There are different types of reinforcement interlayers in terms of material properties, stiffness, and operating principle. In this study, different interlayer typologies are considered regarding stiffness and thickness. In particular, by assuming a reasonable stiffness of 8000 MPa (at 20 °C and 10 Hz load frequency [26]) for the asphalt overlay, the following two categories of reinforcement have been identified.

- Low stiffness reinforcements refer to reinforcements characterized by a modulus of elasticity less than 8000 MPa. This stiffness range is usually associated with the Stress Absorbing Membrane Interlayer (SAMI). Thanks to their low stiffness and high deformability, these layers can absorb stresses and dissipate energy through vertical and horizontal deformation. The SAMI consists of a modified bituminous emulsion covered with a washed and crushed aggregate, forming a diaphragm that distributes the stresses and limits the propagation of cracks from the lower layers of the pavement.
- High stiffness reinforcements refer to reinforcements characterized by a modulus of elasticity greater than 8000 MPa. This reinforcement stiffness can be associated with geomembranes, which are layers of polymer modified bitumen and glass fiber reinforcements, characterized by a waterproofing function [27]. To ensure adhesion between layers, the geomembrane is composed of a three-layer system consisting of a self-adhesive bituminous lower layer, a glass fiber grid middle layer, and a bituminous upper layer with a thermo-adhesive surface.

For the simulations, different values of the elastic modulus are considered for low and high stiffness reinforcements, as shown in Table 1.

Table 1. Reinforcement elastic modules for low stiffness and high stiffness reinforcements.

Reinforcement Stiffness—Elastic Modulus [MPa]	
Low Stiffness Reinforcements -SAMI-	**High Stiffness Reinforcements -Geomembranes-**
400; 500; 600; 700; 800; 900; 1000; 1500; 2000; 3000; 4000; 5000.	10,000; 50,000; 100,000; 200,000.

4.2. Fracture Mechanics

The reflective cracking phenomenon on the pavement is typical of fatigue cracking under repeated load cycles [28,29]. In this study, crack propagation is analyzed by employing Linear Elastic Fracture Mechanics (LEFMs). This assumption is valid for linear elastic material properties. For visco-elastic materials, such as asphalt pavement, specific adjustments should be considered to adapt the LEFMs, as detailed below.

Crack propagation in a pavement can follow two typical paths: bottom-up [30] and top-down [31]. Bottom-up crack propagation occurs when the first crack appears at the bottom of the bonded layers and then propagates upwards, while top-down propagation starts at the surface layer and propagates downwards. Bottom-up propagation is considered for the present analysis, as it is the most common initiation condition caused by tensile stress at the lower surface of the base layer. To describe crack propagation, fracture mechanics should be considered. LEFMs divide the phenomenon into three stages: (i) crack initiation, (ii) crack propagation, and (iii) system failure. In addition, crack propagation may follow a different trend if a reinforcing layer is applied to an already cracked pavement. The crack propagation scheme for an unreinforced and a reinforced pavement is synthesized in Figure 2.

Relevant to this study, each stage of crack propagation is described in detail below.

1. Stage 1: Crack initiation. The crack initiation phase consists of two steps: an initial microcrack and a subsequent macrocrack, corresponding to the formation of a visibly damaged zone. The initiation phase is characterized by the number of repeated load cycles (N_i) causing the appearance of the first damage state within the affected layer. The location of the crack initiation zone depends on the presence of a localized defect.

It is hypothesized that the first crack initiates at the base of the bonded layers and then propagates upwards (bottom-up fatigue cracking), as this crack initiation mechanism is the most common in fatigue cracking [30]. The loading condition causing the crack formation is assumed to be a tensile stress opening the crack, since it has been shown to be the most critical condition for the crack propagation stage as well (Stage 2) [32].

2. Stage 2: Crack propagation. Once the first crack has appeared, if the pavement continues to be cyclically loaded, the crack will continue to propagate. A plastic zone develops at the tip of the crack, the size of which increases as the crack grows [33]. If the plastic region is small enough to be contained within the elastic singularity region, the Stress Intensity Factor (SIF) can be adopted to describe the stress field around the defect [34]. The SIF, represented by the variable K, can be adopted to estimate the fracture dimension (a) during its propagation. The SIF depends on the geometric and loading conditions and can be expressed in general terms for different loading modes by Equation (1).

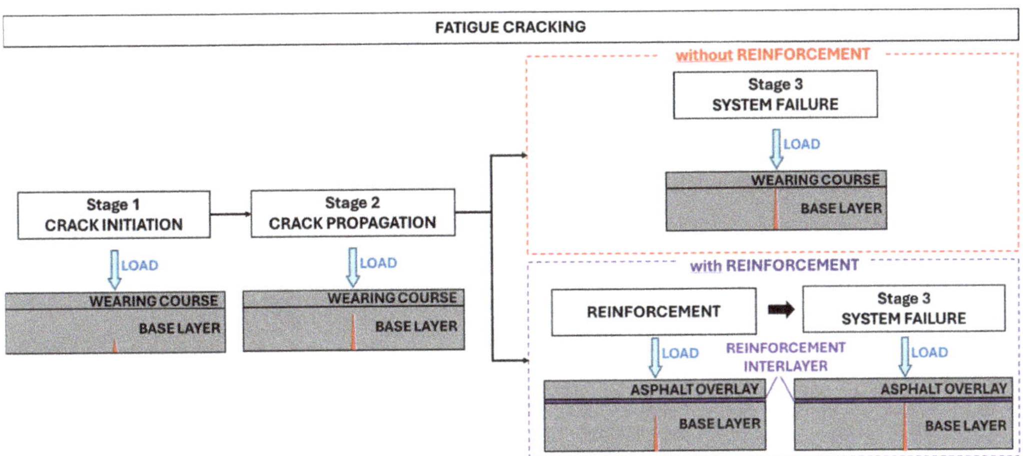

Figure 2. Crack propagation scheme for a non-reinforced and a reinforced pavement.

$$K = Y\sigma\sqrt{\pi a} \tag{1}$$

where σ is the applied stress, a is the crack size, and Y is an adimensional shape factor depending on the geometry and loading mode. To define the SIF, an energetic approach based on the J-integral is adopted. The J-integral is the energy parameter describing the concentration of stress over an area, as given by Equation (2).

$$J = -\frac{dU}{dA} \tag{2}$$

where U is the potential energy, and A is the fracture area. For visco-elastic materials such as asphalt pavement, the J-integral is expressed by Equation (3), through which the SIF is obtained, as expressed in Equation (4).

$$J = \frac{K^2(1-\nu^2)}{E_R} \tag{3}$$

$$K = \sqrt{\frac{E_R \cdot J}{(1-\nu^2)}} \tag{4}$$

where E_R is the reference stiffness modulus for the material, and ν is the Poisson's coefficient. The crack growth rate is a function of the change in the SIF (ΔK). The relationship describing crack propagation is expressed by Equation (5):

$$\frac{da}{dN} = f(\Delta K,\ R) \tag{5}$$

where a is the fracture dimension, N is the number of loading cycles, da/dN is the crack growth per load cycle, ΔK is the change in the SIF, and R is the fracture resistance of the material. Equations (6) and (7) can be adopted for determining the aforementioned parameters.

$$\mathrm{R} = \frac{K_{min}}{K_{max}} \tag{6}$$

$$\Delta K = K_{max} - K_{min} \tag{7}$$

By means of the relationship described in Equation (5), it is possible to estimate the fatigue life of the pavement, which refers to the number of load cycles that will bring the pavement to failure. The number of cycles to failure is calculated as the number of cycles required to propagate the crack from an initial height (a_0) to a final height (a_f), as expressed by Equation (8).

$$N = \int_{a_0}^{a_f} \frac{da}{f(\Delta K,\ R)} \tag{8}$$

In the crack propagation phase, the evolution essentially depends on the material properties. Indeed, the literature has shown that it is possible to characterize the trend of the defect propagation rate as a function of the K parameter using an exponential law that can be linearized in the logarithmic plane. The exponential law describing this trend is known as Paris' law and is expressed in Equation (9), where A and n are parameters characterizing the material.

$$\frac{da}{dN} = A(\Delta K)^n \tag{9}$$

In particular, n is assumed to be equal to 3.4 [35,36], while A is determined through Molenaar theory [37], as shown in Equation (10). This approach allows one to account for the visco-elastic properties of the analyzed material.

$$\log A = 4.389 - 2.52\ \log(E_r \cdot \sigma_t \cdot n) \tag{10}$$

where σ_t is the tensile strength of the material, which is assumed to be equal to 1.8 MPa.

3. Stage 3: System failure. System failure corresponds to the loss of the safety and operability of the road surface. In the case of a reinforced pavement, system failure is reached when the crack has propagated through the full thickness of the base layer reaching the reinforcement layer. For consistency, the system failure for a non-reinforced pavement is assumed to occur when the crack propagates through the entire base layer.

4.3. Method Validation

To validate the method adopted for the solution, an elementary model to describe the pavement is initially considered, namely a single-layer beam. The aim of validation is to confirm the ability of the software considered (COMSOL Multiphysics, and Abaqus) to describe the pavement behavior in terms of the following:

1. Stress trend, which should be consistent with the stress results from the de Saint Venant theory;
2. Crack propagation, which must be in accordance with the results based on Paris' law.

In particular, the validation is performed on a single-layer beam in the absence of a crack for the stress condition and in the presence of a crack for the crack propagation case. Moreover, the following hypotheses have been considered:

- Linear elastic material is assumed to describe asphalt behavior (reasonable assumption in the case of fatigue behavior with a high number of load cycles [38]); the material properties are therefore expressed in terms of the Young's modulus ($E = 8000$ MPa) and Poisson's coefficient ($\nu = 0.3$);
- Plane deformation state ($\varepsilon_x \neq 0, \varepsilon_y \neq 0, \varepsilon_z = 0$);
- Beam dimensions (length L = 500 mm and height h = 50 mm) are assumed to comply with the slender beam condition ($L/h \geq 8$), so that de Saint Venant's theory can be applied;
- A three-point loading condition is needed, with a load of 120 N distributed over an area of 10 mm (distributed load of 12 N/mm).

For the cracked beam simulation scenario, different crack dimensions (a) are assumed for the two software programs:

- For COMSOL Multiphysics, a variable crack height from 2.5 mm to 30 mm with 0.5 mm increments is considered;
- For Abaqus, crack heights of 5, 10, 15, 20, and 25 mm are considered.

Therefore, the single-layer beam scheme considered for stress validation is illustrated in Figure 3a, while the adopted configuration for crack propagation validation is represented in Figure 3b.

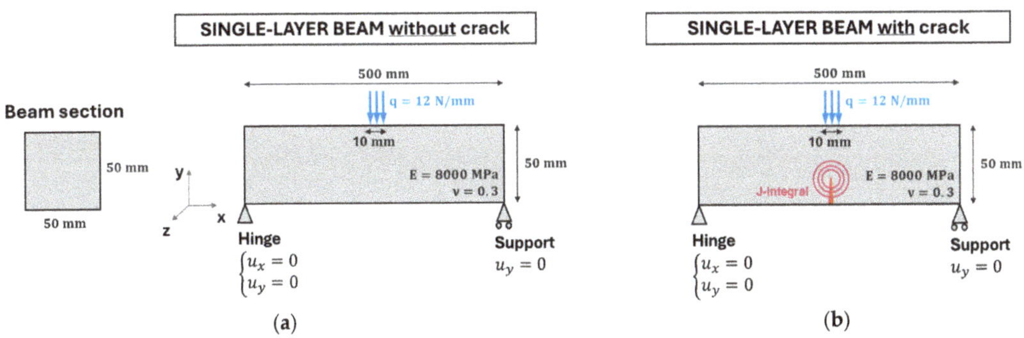

Figure 3. Single-layer beam scheme considered for model validation: (**a**) non-cracked single-layer beam for stress condition validation; (**b**) cracked single-layer beam for crack propagation validation.

4.3.1. Stress Condition

The validation of the stress condition is performed on the single-layer beam without a crack (Figure 3a). In fact, the stress results obtained from the software are compared with those obtained by de Saint Venant's theory. The stresses are calculated at the beam mid-point at various heights. However, as these stresses are influenced by local effects due to the presence of the load at the mid-point, stresses are also calculated at 200 mm from this point. Considering the trend of stresses corresponding to a section slightly shifted from the centerline (at 200 mm), where the concentrated load is applied allows for the effect of stress concentration to be isolated.

The graphical representation of the stresses at the beam centerline (Figure 4a) and 200 mm from the centerline (Figure 4b) shows that there is an almost perfect correspondence between the software and de Saint Venant's theory, except for localized effects caused by the load application. It can be seen that the middle section is affected by stress concentration effects due to the presence of the load, while the section 200 mm away shows an almost perfect correspondence between the software and de Saint Venant's theory. In fact, the stresses assume the pattern of the typical butterfly diagram.

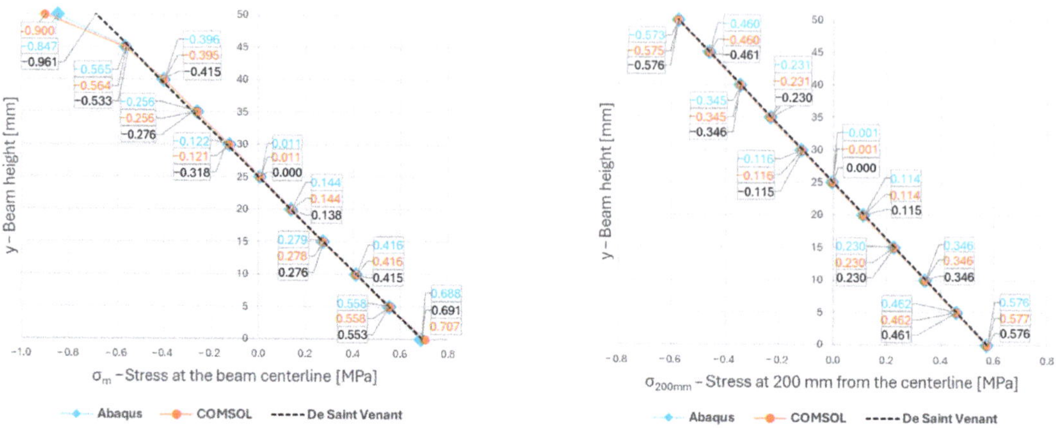

(a) (b)

Figure 4. Comparison of stress diagram in single-layer beam obtained with de Saint Venant theory and software: (**a**) at the beam centerline; (**b**) at 200 mm from the beam centerline.

4.3.2. Crack Propagation

The validation of the crack propagation is performed on the single-layer beam with the crack (Figure 3b). The software results are compared in terms of the Stress Intensity Factor (SIF) with the literature equations for the three-point bending test. In particular, the formulations of Anderson [33] and Carpinteri [39], described by Equations (11) and (12), respectively, are considered.

$$K = \frac{Q\cdot L}{h^{\frac{3}{2}}\cdot B}\cdot g\left(\frac{a}{h}\right) = \frac{Q\cdot L}{h^{\frac{3}{2}}\cdot B}\cdot\frac{3\sqrt{\frac{a}{h}}}{2+\left(1+2\cdot\frac{a}{h}\right)\cdot\left(1-\frac{a}{h}\right)^{\frac{3}{2}}}\cdot\left[1.99-\frac{a}{h}\cdot\left(1-\frac{a}{h}\right)\cdot\left(2.15-3.93\cdot\frac{a}{h}+2.7\cdot\frac{a}{h}^{2}\right)\right] \tag{11}$$

$$K = \frac{Q\cdot L}{h^{\frac{3}{2}}\cdot B}\cdot f\left(\frac{a}{h}\right) = \frac{Q\cdot L}{h^{\frac{3}{2}}\cdot B}\cdot 2.9\cdot\left(\frac{a}{h}\right)^{\frac{1}{2}} - 4.6\cdot\left(\frac{a}{h}\right)^{\frac{3}{2}} + 21.8\cdot\left(\frac{a}{h}\right)^{\frac{5}{2}} - 37.6\cdot\left(\frac{a}{h}\right)^{\frac{7}{2}} + 38.7\cdot\left(\frac{a}{h}\right)^{\frac{9}{2}} \tag{12}$$

where Q is the concentrated load, L is the beam length, h is the beam height, B is the beam thickness, and $\frac{a}{h}$ is the ratio between the crack dimension and the beam height.

The results in terms of the SIF for different crack dimensions are synthesized in Table 2. It can be noted that the SIF shows a good correspondence between the software and literature equations.

Table 2. Comparison of the SIF obtained with the literature equations and software for different crack dimensions.

Crack Dimension	Stress Intensity Factor—K [N/m$^{1.5}$]			
a [mm]	Abaqus	COMSOL	Anderson	Carpinteri
5	91,444	90,577	90,906	89,071
10	129,826	129,317	126,101	125,430
15	169,746	168,887	163,277	163,446
20	220,866	218,773	212,711	212,629
25	291,328	291,290	285,769	286,028

The validation process confirmed the reliability of using both software programs to represent the stress condition and crack propagation. Therefore, only *COMSOL Multiphysics* is referred to in the following, as it allows for the easier modeling of the crack dimension. In particular, the crack is modeled with a variable height ranging from 2.5 mm to 30 mm with 0.5 mm increments.

4.4. Two-Layer System

Once the feasibility of adopting the software has been validated, the assumed model is refined to better represent the actual on-site condition of the pavement. Therefore, a two-layer beam is considered, with the top layer representing the wearing course and the bottom layer representing the base layer. It is assumed that there is perfect adhesion between the two layers. This condition is proven to be achieved in properly constructed pavements [19]. The graphical representation of the two-layer beam is shown in Figure 5. This pavement configuration allows for the investigation of the effect of different fracture dimensions on the stress condition and crack propagation of the two-layer system. Therefore, simulations for the two-layer beam can be extrapolated to the behavior of a non-reinforced pavement.

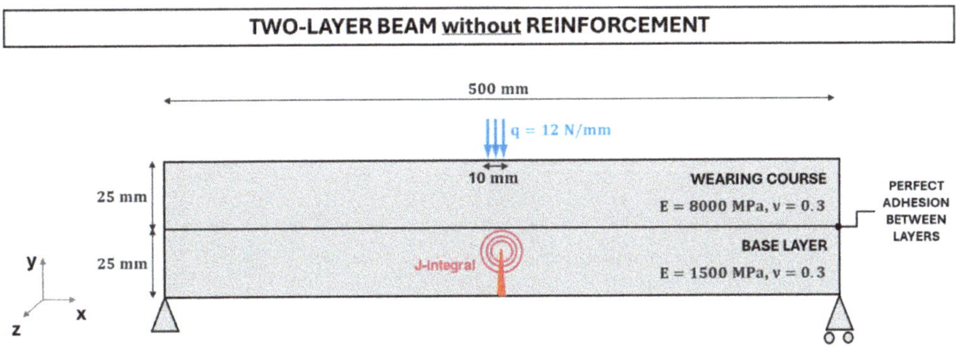

Figure 5. Non-reinforced pavement represented by a two-layer beam scheme.

The following results obtained from the software simulations are analyzed.

1. σ_x: stress along the x-direction calculated at the beam centerline for the entire beam height. Based on Figure 6, it can be noted that when considering increasing crack dimensions (a), an increasing stress trend is registered along the beam centerline. In addition, the maximum tensile stress is recorded in correspondence with the crack tip. Moreover, larger crack heights increase the stress in the wearing course, increasing the likelihood of crack initiation. If the crack extends through the entire base layer (a = 25 mm), the stress in the bottom layer is null and the stress in the top layer is maximized.
2. The J-integral and SIF for different crack dimensions and different radii of the circular integration area (r). A different J-integral will be calculated according to the selected circular radius for the integration area, as shown in Figure 7a. In particular, the value of the J-integral depends on the radius of the circular integration area (r) when the integration surface intersects the boundary between the lower and upper layers. Consequently, a specific SIF is obtained, as illustrated in Figure 7b.
3. N_{nr}: the number of load cycles to failure without reinforcement, where failure is reached when the crack propagates through the entire base layer. The number of cycles leading to system failure is calculated as the cumulative number of load cycles for several increasing crack heights. Thus, according to results expressed in Figure 8, the total number of load cycles that will cause the system to fail without reinforcement is approximately $N_{nr} = 6.71 \cdot 10^8$.

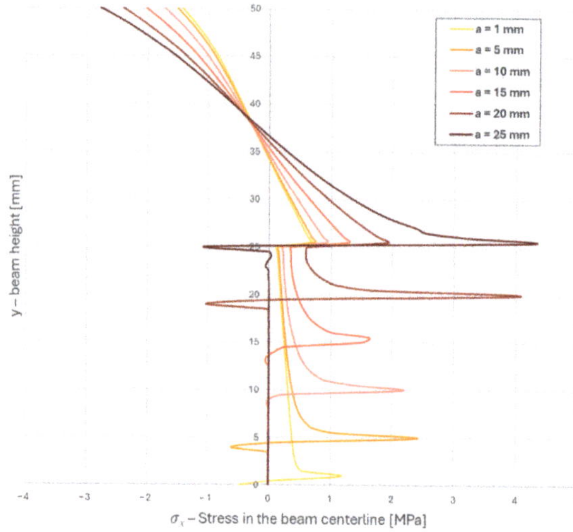

Figure 6. Stress at the beam centerline in a two-layer system for different crack dimensions (*a*).

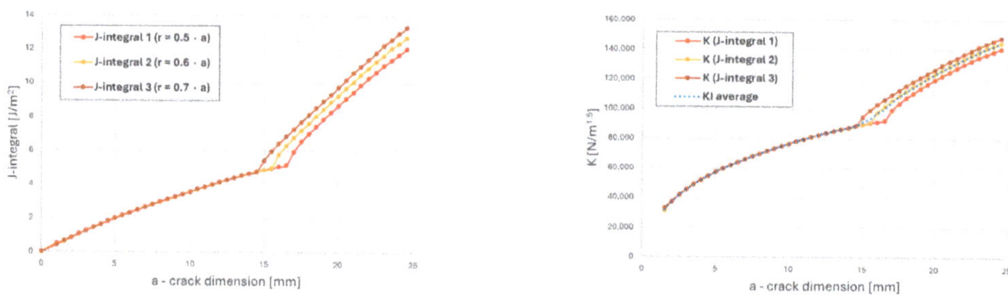

(**a**) (**b**)

Figure 7. Paris' law parameters in a two-layer system for different crack dimensions: (**a**) J-integral; (**b**) Stress Intensity Factor.

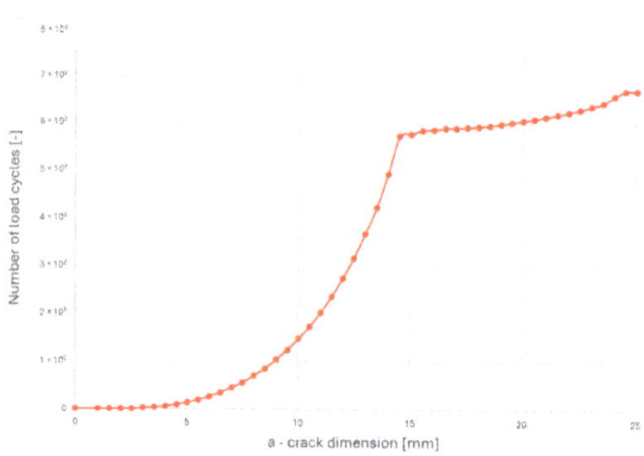

Figure 8. Cumulative load cycles leading the two-layer beam system to failure.

4.5. Three-Layer System

To simulate the effect of reinforcement with reference to the pavement scheme considered in Section 4.4, a three-layer beam is considered. Therefore, the pavement scheme considered is shown in Figure 9. It should be noted that the effect of the reinforcement depends on its thickness and stiffness; thus, different simulations are performed for different reinforcement characteristics.

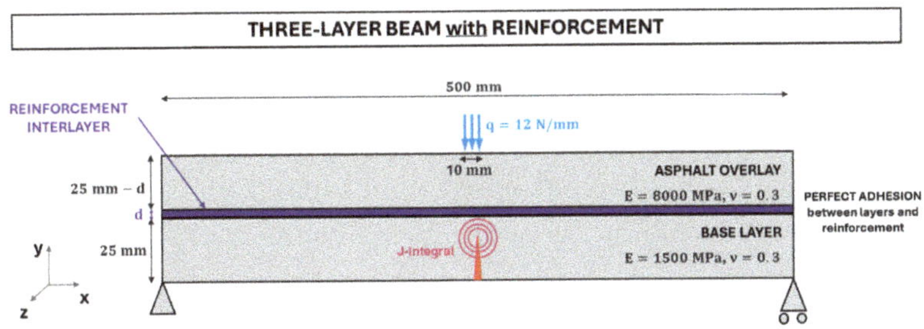

Figure 9. Reinforced pavement represented by a three-layer beam scheme.

Software simulations allow the number of cycles to failure to be determined for different reinforcement thicknesses and stiffnesses. In fact, the output data obtained from the software are the same as those considered in Section 4.4. However, for the reinforced pavement, the number of load cycles that bring the system to failure (crack propagation affecting the whole base layer) is expressed through the symbol N_r.

To determine the effect of the reinforcement on the wearing course, the maximum tensile stress recorded in the surface layer is evaluated for different reinforcement thicknesses and stiffnesses. The results reported in Figure 10 show that the tensile stress in the wearing course increases with the crack dimension. In fact, as the crack propagates from the deeper layer (base) upwards, the maximum stress in the surface layer increases. In addition, the positive effect of the reinforcement in containing the tensile stress in the wearing course is more pronounced at greater thicknesses and stiffnesses. When comparing two different reinforcing layer thicknesses, 0.15 mm (Figure 10a) and 2 mm (Figure 10b), for the same stiffness, the reinforcement proves to be more effective with a greater thickness. This means that the reinforcement layer should be sufficiently stiff and thick to prevent cracking in the wearing course.

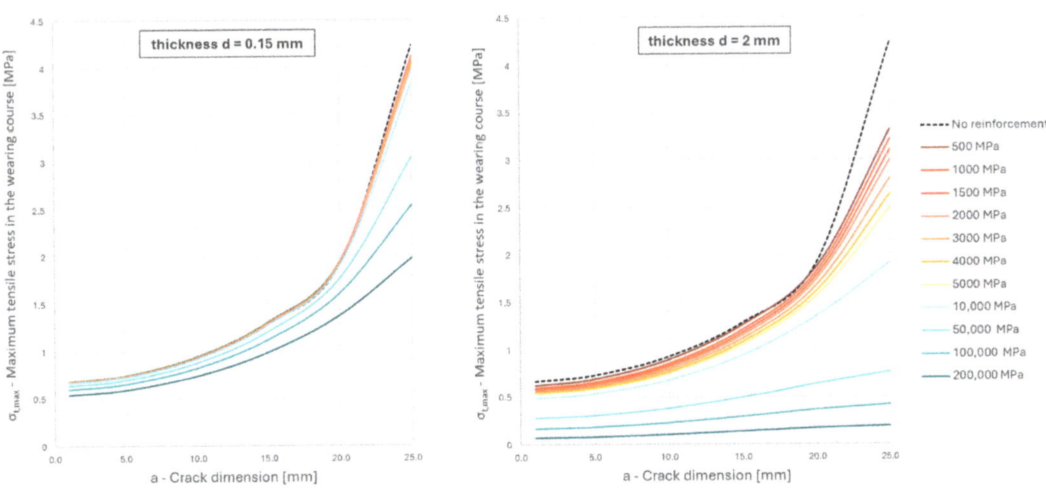

(a) (b)

Figure 10. Maximum tensile stress in the wearing course for increasing crack dimension: (**a**) for a reinforcement thickness d = 0.15 mm; (**b**) for a reinforcement thickness d = 2 mm.

Similarly, when comparing the effect of reinforcement for two different crack dimensions, 1 mm (Figure 11a) and 25 mm (Figure 11b), a combined effect of the layer thickness and stiffness is registered. In small crack dimensions, the effect of the reinforcement layer is significant only for high reinforcement stiffness (>8000 MPa). Conversely, for larger crack dimensions, the effect of reinforcement is noticeable even for lower stiffness layers when they have a sufficient thickness.

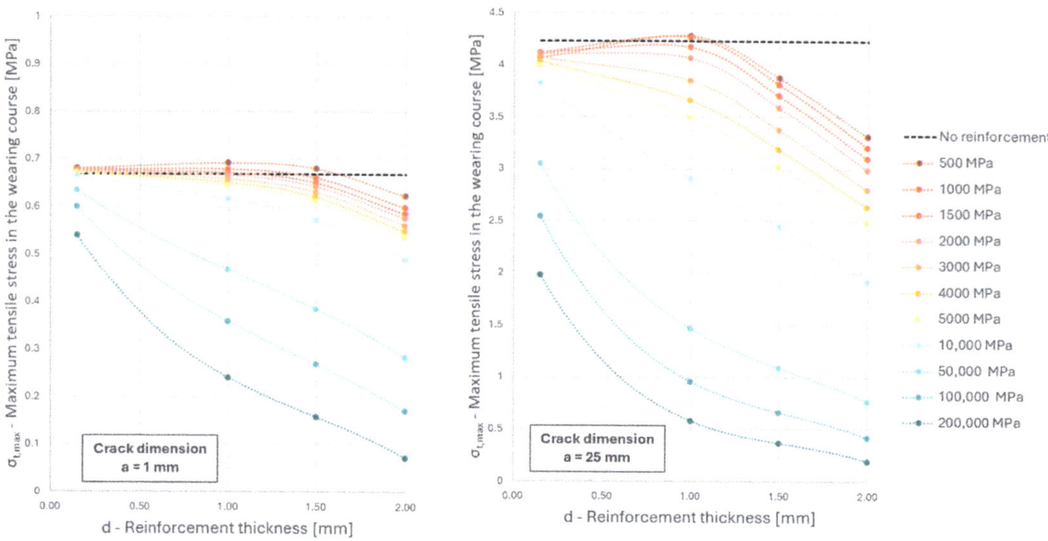

(a) (b)

Figure 11. Maximum tensile stress in the wearing course for increasing reinforcement thickness: (**a**) for a crack dimension a = 1 mm; (**b**) for a crack dimension a = 25 mm.

4.6. Reinforcement Performance

To evaluate the effect of the reinforcement layer on the base layer, crack propagation should be considered. In particular, the performance of the reinforcement is described in terms of the number of cycles to failure with and without reinforcement, through the following Equation (13):

$$\eta = \frac{N_r}{N_{nr}} = \frac{\int_0^H \cdot \frac{da}{(\Delta K_r)^n}}{\int_0^H \cdot \frac{da}{(\Delta K_{nr})^n}} \tag{13}$$

where η is the reinforcement performance, N_{nr} is the number of load cycles to failure without reinforcement, N_r is the number of load cycles to failure with reinforcement, ΔK_{nr} is the SIF for the non-reinforced pavement, and ΔK_r is the SIF for the reinforced pavement. Whenever the performance value is greater than one, the reinforcement has a positive effect on crack propagation in the base layer.

The results in terms of reinforcement performance for different thicknesses and stiffnesses are synthetized in Table 3.

Table 3. Reinforcement performance for different thicknesses and stiffnesses.

Reinforcement Thickness d [mm]	Reinforcement Performance—η [-]																
	Low Stiffness [MPa]											[MPa]		High Stiffness [MPa]			
	400	500	600	700	800	900	1000	1500	2000	3000	4000	5000	8000	10,000	50,000	100,000	200,000
0.15	0.85	0.87	0.88	0.89	0.90	0.90	0.91	0.92	0.93	0.95	0.96	0.97	1.00	1.01	2.04	3.40	34.67
1.00	4.41	1.74	1.19	0.99	0.90	0.85	0.82	0.77	0.76	0.78	0.82	0.86	1.00	1.11	6.90	82.85	632,942.53
1.50	1.50	1.11	0.95	0.87	0.82	0.79	0.76	0.72	0.71	0.73	0.77	0.82	1.00	1.16	17.86	1903.29	6,065,251.51
2.00	0.81	0.72	0.68	0.66	0.64	0.63	0.62	0.61	0.62	0.66	0.71	0.77	1.00	1.20	48.02	2659.01	35,576.08

The results show that high modulus reinforcements consistently exert a positive effect on crack propagation in the base layer due to their stiffening effect. Specifically, they offer increasingly significant benefits with a higher modulus and thickness. Conversely, low stiffness modulus reinforcements have no stiffening effect and, due to stress redistribution, increase the tensile stresses in the base layer and reduce the number of cycles to failure. In fact, for low stiffness reinforcements, the performance η is generally less than one, which represents a negative effect on crack propagation. The only low stiffness reinforcement configuration associated with a positive effect is for extremely low stiffness moduli (<700 MPa) and for intermediate thicknesses (d = 1 mm or d = 1.5 mm).

Figure 12 synthetizes the reinforcement performance at different stiffness levels, confirming that high modulus reinforcements always have a positive effect in delaying crack propagation, while low modulus reinforcements are mainly inefficient except at very low stiffness values.

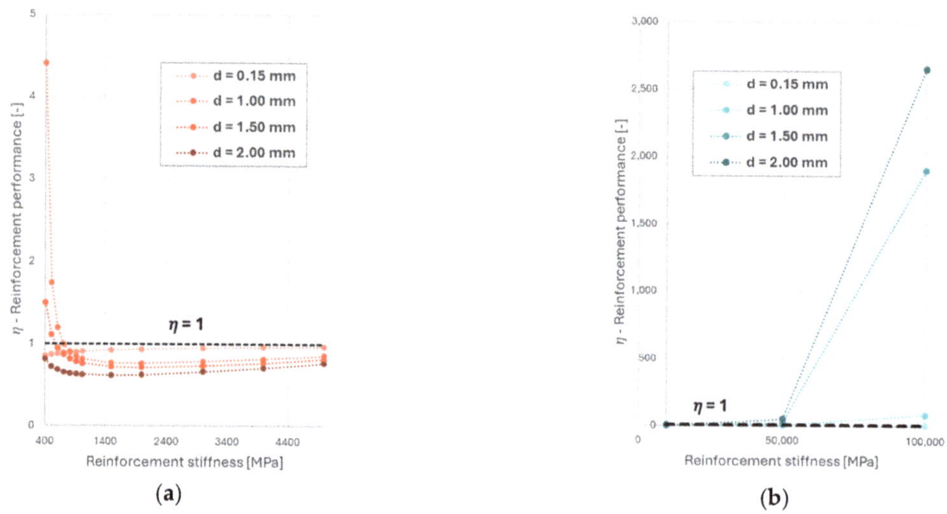

Figure 12. Reinforcement performance at different thicknesses and stiffnesses for: (**a**) low stiffness reinforcements; (**b**) high stiffness reinforcements.

4.7. Multilayer System

Once the main trends between reinforcement thickness/stiffness and its performance have been defined, a more realistic pavement configuration is simulated. In particular, a multilayer system representing an asphalt pavement is considered. Specifically, the asphalt bonded layers (wearing course and base) are placed on top of an unbonded layer (sub-base) laying on the subgrade. The specific pavement scheme without reinforcement and its relative material properties and boundary conditions is shown in Figure 13. In particular, the bottom of the subgrade is considered fixed (displacements are not allowed along both the x and y directions). Lateral boundaries of the multilayer system are represented as a trolley, which prohibits any movement along x and allows translations only along y. The interaction between the different layers is simulated according to two different schemes: (i) a condition of perfect adhesion is assumed at the interface between the bonded layers, i.e., total cooperation with the transfer of vertical stresses and the same horizontal deformations, and (ii) at the interface with at least one unbonded layer, the total absence of adhesion is assumed, with the layers being completely independent from the tensile-deformation point of view, thus eliminating the transfer of horizontal forces. To simulate reflecting cracking, a variable crack dimension between 0.1 mm and 170 mm in 0.5 mm increments is assumed so that for the maximum crack height, the entire base layer is cracked. For the purpose of finite element modeling, the mesh characteristics were determined. A structured triangular mesh with a maximum dimension of 5 mm and densification near the fracture apex was assumed, as detailed in Figure 13.

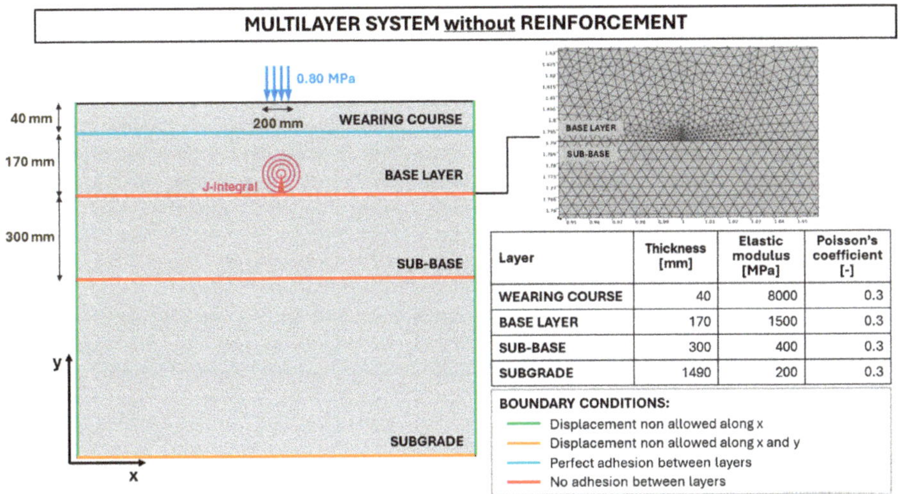

Layer	Thickness [mm]	Elastic modulus [MPa]	Poisson's coefficient [-]
WEARING COURSE	40	8000	0.3
BASE LAYER	170	1500	0.3
SUB-BASE	300	400	0.3
SUBGRADE	1490	200	0.3

Figure 13. Multilayer pavement scheme without reinforcement.

To calculate the J-integral, four different circular integration surfaces with an increasing radius r (10 mm, 20 mm, 30 mm, and 40 mm) are considered. The software, therefore, simulates the stress condition and crack propagation to failure.

Similarly, the reinforced pavement scheme is assumed, as shown in Figure 14. In this case, the software is used to run simulations for different reinforcement thicknesses and stiffnesses. It should be noted that for low stiffness reinforcements (SAMIs), a thickness of 5 mm is taken into account, while for high stiffness reinforcements (geomembranes), a thickness of 0.15 mm is considered. In particular, since the geomembrane is a three-layer system composed of two bituminous layers comprising a fiberglass grid, an equivalent thickness is taken into account. This allows for the consideration of the average reinforcement material properties for software simulation purposes.

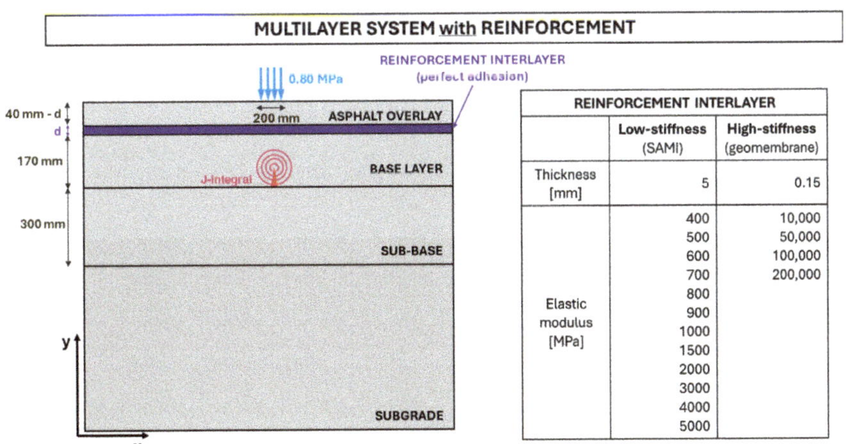

REINFORCEMENT INTERLAYER		
	Low-stiffness (SAMI)	High-stiffness (geomembrane)
Thickness [mm]	5	0.15
Elastic modulus [MPa]	400 500 600 700 800 900 1000 1500 2000 3000 4000 5000	10,000 50,000 100,000 200,000

Figure 14. Multilayer pavement scheme with reinforcement.

The thickness of the asphalt overlay depends on the thickness of the reinforcement, ensuring that the total thickness of reinforcement and overlay is equal to 40 mm. Regarding the elastic modulus, various values have been considered for both low and high stiffness reinforcements, as reported in Figure 14.

5. Results and Discussion

The effect of the reinforcement on the schematized pavement is evaluated by comparing the non-reinforced configuration (Figure 13) with the reinforced one (Figure 14). In particular, two aspects should be considered: the effect on the stress condition in the wearing course, and the crack propagation in the base layer. In fact, the reinforcement should be able to both prevent new cracking in the wearing course and delay crack propagation into a deeper layer of the pavement.

5.1. Reinforcement Effect on the Stress Condition of the Wearing Course

To evaluate the effectiveness of the reinforcement in preventing cracking in the wearing course, the maximum tensile stress is considered for various scenarios. In particular, the unreinforced pavement is compared with the reinforced pavement for various stiffness values.

5.1.1. Low Stiffness Modulus Reinforcements—SAMI

The results of the simulations in terms of the maximum tensile stress in the wearing course for low stiffness reinforcements are shown in Figure 15. It can be noted that low stiffness reinforcements (E < 8000 MPa) with a thickness of 5 mm are not always beneficial for preventing cracking in the wearing course. In fact, extremely low stiffness reinforcements (E < 700 MPa) induce a tensile stress in the wearing course that is even more critical compared to the unreinforced pavement. This phenomenon may be due to the presence of an interlayer contributing to the transfer of stresses from the lower cracked layers to the new asphalt overlay, overloading it. Only for fairly large cracks (a > 150 mm), which nearly affect the entire base layer, a slight positive effect of the reinforcement is observed. Conversely, stiffness values equal or greater than 700 MPa allow a positive effect to be achieved in the wearing course. In addition, the benefit is more pronounced for increasing stiffness. However, for the simulated scenarios, any reinforced pavement will still reach a fracture state if the crack extends to the full height of the deeper layers, as the tensile strength (1.80 MPa) is overcome.

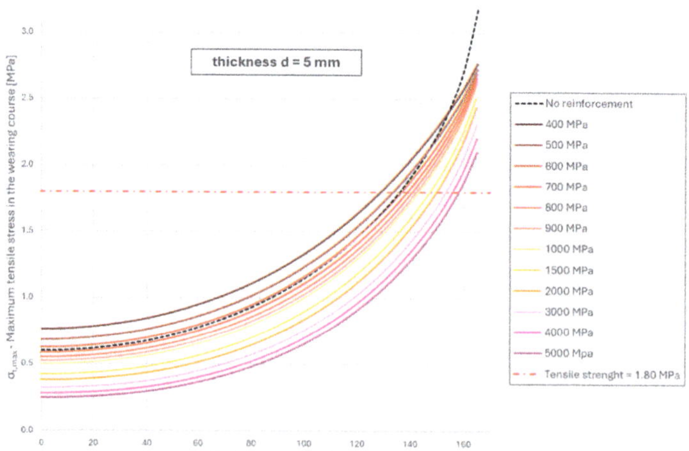

Figure 15. Maximum tensile stress in the wearing course for low stiffness reinforcements.

5.1.2. High Stiffness Modulus Reinforcements—Geomembranes

Regarding high stiffness reinforcements (E > 8000 MPa), a constant thickness of 0.15 mm is considered. As shown in Figure 16, high stiffness reinforcements are almost always effective in preventing cracking in the wearing course. In fact, even for a crack dimension of 170 mm affecting the entire base layer, the maximum tensile stress in the wearing course is below the tensile strength (equal to 1.8 MPa). The tensile strength is exceeded for a crack dimension of 170 mm only in the case of a reinforcement stiffness of 10,000 MPa, meaning that cracking will occur in the wearing course.

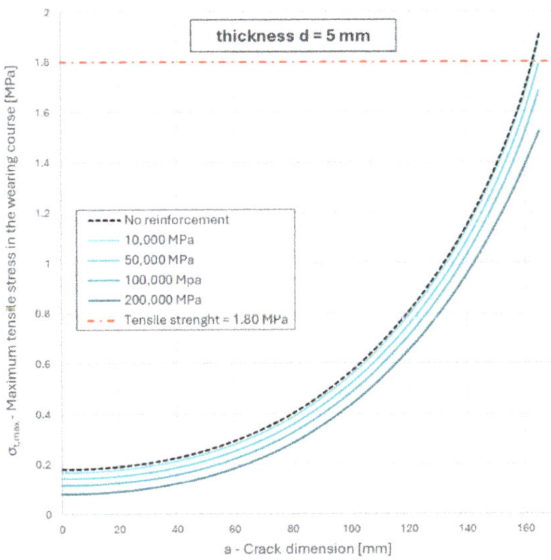

Figure 16. Maximum tensile stress in the wearing course for high stiffness reinforcements.

Therefore, high stiffness reinforcements with a modulus of elasticity greater or equal to 50,000 MPa are effective in preventing cracking in the surface layer. In addition, the maximum tensile stress in the wearing course is reduced as the stiffness of the reinforcement layer increases.

5.2. Reinforcement Effect on the Cracking State of the Base Layer

To evaluate the effectiveness of the reinforcement in delaying crack propagation in the base layer, the number of load cycles to failure are considered for various scenarios. Specifically, the unreinforced pavement is compared with the reinforced one for various stiffness values. Thus, the reinforcement performance (η) is computed according to Equation (13). Performance values greater than one express a positive effect of the reinforcement, as the number of cycles to failure is increased compared to the unreinforced pavement. Conversely, performance values lower than one indicate a detrimental effect of the reinforcement on crack propagation in the base layer.

5.2.1. Low Stiffness Modulus Reinforcements—SAMI

Performance values for low stiffness reinforcements are calculated by comparing the number of cycles to failure with no reinforcement and with 5 mm of reinforcement. The results presented in Figure 17 show that a beneficial effect is registered for a reinforcement stiffness of less than 900 MPa. In fact, for this condition, a performance value greater than one is obtained, meaning that the reinforcement can counteract crack propagation in the base layer. This means that low stiffness reinforcements, due to their high deformability, can slow down crack propagation. Conversely, as stiffness increases, the positive effect of stress absorption is gradually reduced, so that for a modulus of elasticity between 900 MPa and 5000 MPa, the reinforcement is neither deformable enough to absorb stresses nor stiff enough to reinforce the deeper layers. As a result, the performance values are less than one and the reinforcement is not beneficial.

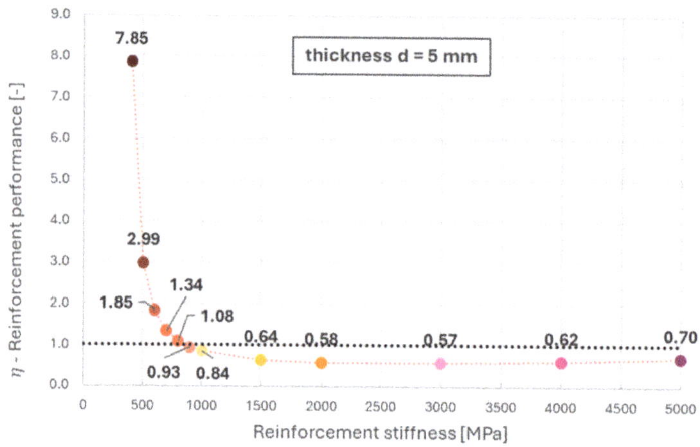

Figure 17. Reinforcement performance for low stiffness reinforcements.

5.2.2. High Stiffness Modulus Reinforcements—Geomembranes

Performance values for high stiffness reinforcements are calculated by comparing the number of cycles to failure with no reinforcement and with 0.15 mm of reinforcement. The results presented in Figure 18 show that a beneficial effect is always registered. In fact, for a modulus of elasticity greater or equal to 10,000 MPa, the performance values are always greater than one. This means that the number of cycles to failure is always increased compared to the unreinforced pavement. In the case of high stiffness reinforcements, the beneficial effect on the base layer in terms of crack propagation delay increases with stiffness.

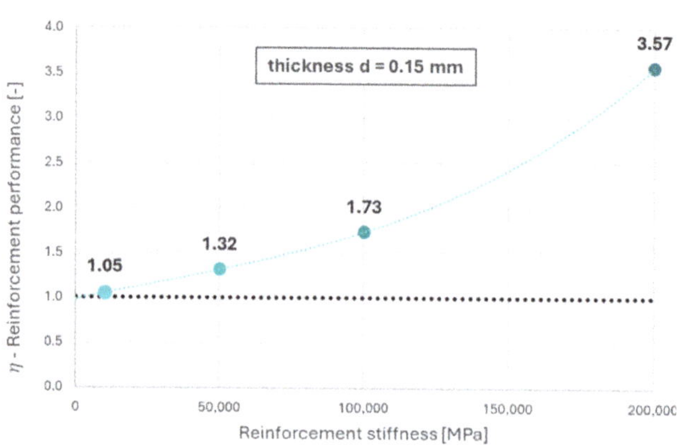

Figure 18. Reinforcement performance for high stiffness reinforcements.

6. Conclusions

This study aimed to evaluate the effectiveness of reinforcement interlayers in road pavement rehabilitation by mitigating fatigue cracking. Two types of interlayers were considered: (i) low stiffness reinforcements (SAMIs) and (ii) high stiffness reinforcements (geomembranes). The benefits of reinforcement were evaluated based on the stress condition of the wearing course and the cracking state of the base layer.

The results showed that geomembranes always have a positive effect in terms of tensile stress in the wearing course. Moreover, interlayers with a very high stiffness (≥50,000 MPa) are effective in preventing cracking in the surface course, even with a fully cracked deeper course. In addition, geomembranes can delay crack propagation as the performance values are always greater than one. The performance increases with increasing stiffness, ranging from 1.05 (for 10,000 MPa reinforcement stiffness) to 3.57 (for 200,000 MPa reinforcement stiffness). For example, a 50,000 MPa geomembrane (0.15 mm thick) can increase the service life of a pavement by approximately 30%.

While geomembranes always have a positive effect, SAMIs do not always have a positive effect. In terms of tensile stress in the wearing course, a stiffening effect is required and a sufficiently high stiffnesses is necessary to have a positive effect on the existing pavement (≥700 MPa). In addition, the tensile strength in the wearing course is always exceeded for a totally cracked base layer, meaning that SAMIs are not fully effective in preventing cracking. In terms of crack propagation in the base layer, SAMIs only have a positive effect for sufficiently low stiffness (sufficiently deformable), showing performance values greater than one for reduced stiffness (≤800 MPa). Therefore, SAMIs present some critical issues related to the fact that they should be sufficiently deformable to delay crack propagation in the base layer, but stiff enough to prevent cracking in the wearing course.

In conclusion, geomembranes always have a positive effect both in delaying crack propagation in the deeper layers and in preventing cracking in the surface layer. On the contrary, SAMIs have a positive effect in reducing the probability of cracking in the wearing course for sufficiently high stiffnesses (≥700 MPa) and in retarding the propagation of cracks for sufficiently low stiffnesses (≤800 MPa). Thus, SAMIs could be effective for intermediate stiffnesses (700/800 MPa). However, the effect would be slightly positive and not comparable to that of geomembranes.

The results of this study, based on the simulation of pavement behavior, could be useful to plan an experimental analysis to confirm the results obtained, as the focus of a further agenda.

Author Contributions: Conceptualization, M.C.; methodology, E.T.; software, A.A.; validation, A.A.; formal analysis, A.A.; investigation, G.R. and C.S.; resources, G.R. and C.S.; data curation, G.R. and C.S.; writing—original draft preparation, A.A.; writing—review and editing, E.T.; visualization, A.A.; supervision, M.C. and E.T. All authors have read and agreed to the published version of the manuscript.

Funding: This research received no external funding.

Data Availability Statement: Dataset available on request from the authors.

Conflicts of Interest: The authors declare no conflicts of interest.

References

1. Walubita, L.F.; Simate, G.S.; Ofori-Abebresse, E.; Martin, A.E.; Lytton, R.L.; Sanabria, L.E. Mathematical formulation of HMA crack initiation and crack propagation models based on continuum fracture-mechanics and work-potential theory. *Int. J. Fatigue* **2012**, *40*, 112–119. [CrossRef]
2. Harne, V.R.; Tripathi, R.K.; Guzzarlapudi, S.D. Effect of extrinsic and intrinsic parameters on fatigue performance of hot mix asphalt mixtures: A review. *J. Build. Pathol. Rehabil.* **2024**, *9*, 53. [CrossRef]
3. Alnaqbi, A.J.; Zeiada, W.; Al-Khateeb, G.; Abttan, A.; Abuzwidah, M. Predictive models for flexible pavement fatigue cracking based on machine learning. *Transp. Eng.* **2024**, *16*, 100243. [CrossRef]
4. Rith, M.; Woo Lee, S. Evaluation of Asphalt Overlay Pretreatments against Reflective Crack Using Association Rule Mining. *J. Transp. Eng. Part B Pavements* **2021**, *147*, 04021042. [CrossRef]
5. Raheem, H.M.; Abduljabbar, A.S. Using Deferment Improvement Techniques to Mitigate Reflection Cracks in Composite Pavements. *AIP Conf. Proc.* **2023**, *2775*, 060016. [CrossRef]
6. Zvonarić, M.; Dimter, S. Prevention and remediation measures for reflective cracks in flexible pavements. *Gradjevinar* **2022**, *74*, 189–197. [CrossRef]

7. Asadi, S.; Shafabakhsh, G. Presenting a Statistical Model of Fatigue Prediction for the Effect of Loading Frequency on Reflective Cracks Propagation on Asphalt Layers Improved by Geosynthetics. *J. Rehabil. Civ. Eng.* **2024**, *12*, 18–33. [CrossRef]
8. Sudarsanan, N.; Mohapatra, S.R.; Karpurapu, R.; Amirthalingam, V. Use of Natural Geotextiles to Retard Reflection Cracking in Highway Pavements. *J. Mater. Civ. Eng.* **2018**, *30*, 04018036. [CrossRef]
9. Nejad, F.M.; Noory, A.; Toolabi, S.; Fallah, S. Effect of using geosynthetics on reflective crack prevention. *Int. J. Pavement Eng.* **2015**, *16*, 477–487. [CrossRef]
10. Sun, H. geng Development of an indoor test method for evaluating the anti-reflection crack performance of asphalt overlay. *Case Stud. Constr. Mater.* **2023**, *19*, e02241. [CrossRef]
11. Moses, O. Mechanical Behaviour of Stress Absorbing Membrane Interlayers. Ph.D. Thesis, The University of Nottingham, Nottingham, UK, 2012.
12. Ogundipe, O.M.; Thom, N.H.; Collop, A.C. Evaluation of performance of stress-absorbing membrane interlayer (SAMI) using accelerated pavement testing. *Int. J. Pavement Eng.* **2013**, *14*, 569–578. [CrossRef]
13. Ogundipe, O.M.; Thom, N.; Collop, A. Investigation of crack resistance potential of stress absorbing membrane interlayers (SAMIs) under traffic loading. *Constr. Build. Mater.* **2013**, *38*, 658–666. [CrossRef]
14. Ogundipe, O.M.; Thom, N.H.; Collop, A.C. Finite element analysis of overlay incorporating stress absorbing membrane interlayers against reflective cracking. *J. Mod. Transp.* **2014**, *22*, 104–111. [CrossRef]
15. Baek, C. Performance evaluation of fiber-reinforced, stress relief asphalt layers to suppress reflective cracks. *Appl. Sci.* **2020**, *10*, 7701. [CrossRef]
16. Zhang, K.; Zhang, Z.; Luo, Y. Material Composition Design and Anticracking Performance Evaluation of Asphalt Rubber Stress-Absorbing Membrane Interlayer (AR-SAMI). *Adv. Mater. Sci. Eng.* **2018**, *2018*, 8560604. [CrossRef]
17. Yuya, W.; Yasushi, T.; Futoshi, K.; Kazuhiro, W. Evaluation of the Effect of Interlayer Bonding Condition on the Deterioration of Asphalt Pavement. *Transp. Res. Rec.* **2023**, *2677*, 500–508. [CrossRef]
18. Sudarsanan, N.; Karpurapu, R.; Amrithalingam, V. Critical review on the bond strength of geosynthetic interlayer systems in asphalt overlays. *Jpn. Geotech. Soc. Spec. Publ.* **2015**, *2*, 2296–2301. [CrossRef]
19. Canestrari, F.; Cardone, F.; Gaudenzi, E.; Chiola, D.; Gasbarro, N.; Ferrotti, G. Interlayer bonding characterization of interfaces reinforced with geocomposites in field applications. *Geotext. Geomembr.* **2022**, *50*, 154–162. [CrossRef]
20. Raab, C.; Arraigada, M.; Partl, M.N.; Schiffmann, F. Cracking and interlayer bonding performance of reinforced asphalt pavements. *Eur. J. Environ. Civ. Eng.* **2017**, *21*, 14–26. [CrossRef]
21. Zhu, Z.; Xiao, P.; Kang, A.; Kou, C.; Wu, B.; Ren, Z. Innovative design of self-adhesive basalt fiber mesh geotextiles for enhanced pavement crack resistance. *Geotext. Geomembr.* **2023**, *52*, 368–382. [CrossRef]
22. Elseifi, M.A.; Baek, J.; Dhakal, N. Review of modelling crack initiation and propagation in flexible pavements using the finite element method. *Int. J. Pavement Eng.* **2018**, *19*, 251–263. [CrossRef]
23. Zhang, Z.; Roque, R.; Birgisson, B. Evaluation of laboratory-measured crack growth rate for asphalt mixtures. *Transp. Res. Rec.* **2001**, *1767*, 67–75. [CrossRef]
24. Luo, X.; Zhang, Y.; Lytton, R.L. Implementation of pseudo J-integral based Paris' law for fatigue cracking in asphalt mixtures and pavements. *Mater. Struct. Constr.* **2016**, *49*, 3713–3732. [CrossRef]
25. Santos, I.; Chupin, O.; Piau, J.M.; Marsac, P.; Hammoum, F. Modeling of fatigue cracking in asphalt materials based on the paris law with an initialization term-calibration from notched and unnotched specimens. *J. Test. Eval.* **2023**, *51*, 2132–2146. [CrossRef]
26. Brovelli, C.; Crispino, M.; Pais, J.; Pereira, P. Using polymers to improve the rutting resistance of asphalt concrete. *Constr. Build. Mater.* **2015**, *77*, 117–123. [CrossRef]
27. International Geosynthetics Society. Available online: https://www.geosyntheticssociety.org/educational-documents/geosynthetics-classification/ (accessed on 1 May 2024).
28. Chen, A.; Airey, G.D.; Thom, N.; Li, Y. Characterisation of fatigue damage in asphalt mixtures using X-ray computed tomography. *Road Mater. Pavement Des.* **2023**, *24*, 653–671. [CrossRef]
29. Si, C.; Cao, H.; Fan, T.; Jia, Y.; Wang, X.; Li, S.; Xu, Z.; Gu, J. Study on crack propagation behavior of bridge deck asphalt pavement. *Constr. Build. Mater.* **2024**, *425*, 136136. [CrossRef]
30. Zhang, J.; Zhang, J.; Cao, D.; Ding, Y.; Zhou, W. Mechanistic analysis of bottom-up crack in asphalt pavement using cohesive zone model. *Theor. Appl. Fract. Mech.* **2023**, *125*, 103904. [CrossRef]
31. Sun, L.; Wang, G.; Zhang, H.; Liu, L. Initiation and propagation of top-down cracking in asphalt pavement. *Appl. Sci.* **2018**, *8*, 774. [CrossRef]
32. Casey, D.B.; Collop, A.C.; Grenfell, J.R.; Airey, G.D. Stress Intensity Factors at the Tip of a Surface Initiated Crack Caused by Different Contact Pressure Distributions. *Procedia-Soc. Behav. Sci.* **2012**, *48*, 733–742. [CrossRef]
33. Ted, L. *Anderson Fracture Mechanics: Fundamentals and Applications*, 3rd ed.; CRC Press: Boca Raton, FL, USA, 2005; ISBN 9780429125676.
34. Irwin, G.R. Analysis of Stresses and Strains Near the End of a Crack Traversing a Plate. *J. Appl. Mech.* **1957**, *24*, 361–364. [CrossRef]
35. Elseifi, M. Performance Quantification of Interlayer Systems in Flexible Pavements Using Finite Element Analysis, Instrument Response, and Non Destructive Testing. Ph.D. Thesis, Virginia Tech, Blacksburg, VA, USA, 2003.
36. Ferraresi, F. Analisi Numerico-Sperimentale Sulla Resistenza a Fatica dei Conglomerati Bituminosi Rinforzati. Master's Thesis, Politecnico di Milano, Milan, Italy, 2010.

37. Francken, L.; Beuving, E.; Molenaar, A.A.A. *Reflective Cracking in Pavements: Design and Performance of Overlay Systems;* CRC Press: Boca Raton, FL, USA, 1996; Volume 1, p. 576.
38. Taliercio, A. Capitolo 7—Il Problema di Saint-Venant. In *Introduzione alla Meccanica dei Solidi*, 2nd ed.; Società Editrice Esculapio: Bologna, Italy, 2014.
39. Carpinteri, A. *Meccanica dei Materiali e della Frattura;* Pitagora Editrice: Bologna, Italy, 1992.

buildings

MDPI

Article

Modeling and Laboratory Investigation of Tack Coats as Bituminous Pavement Interlayer

Alessandro Steffanoni [1], Michel Di Tommaso [2], Vito Giovanni Gallo [2], Giuseppe Macaluso [2], Carmine Rizzato [2], Misagh Ketabdari [1] and Emanuele Toraldo [1,*]

[1] Department of Civil and Environmental Engineering, Politecnico di Milano, Piazza Leonardo da Vinci 32, 20133 Milan, Italy; alessandro.steffanoni@mail.polimi.it (A.S.); misagh.ketabdari@polimi.it (M.K.)

[2] IMM SA—Istituto Meccanica dei Materiali SA, Via al Molino 55, CH 6926 Montagnola, Switzerland; ditommaso@imm.ch (M.D.T.); gallo@imm.ch (V.G.G.); macaluso@imm.ch (G.M.); rizzato@imm.ch (C.R.)

* Correspondence: emanuele.toraldo@polimi.it

Abstract: The adhesive properties of tack coats between asphalt pavement layers are crucial for the pavement's structural behavior. This study first involved numerical analyses to compare stress patterns, deformations, and displacements in the pavement structure under various geometric and mechanical conditions. A rational calculation method based on the theory of elastic multilayer systems was used to quantify the impact of layer properties such as thickness, stiffness modulus, and Poisson's ratio on interlayer bonding. Three bonding conditions—Full Friction, Partial Bonding, and Full Debonding—were analyzed to understand the tack coat's effect between the top two layers. The second phase involved characterizing the mechanical behavior of the interface through shear strength tests (Leutner shear test) on both laboratory-prepared specimens and samples from a 10-year-old highway. Specimens were prepared using a Roller Compactor and tested under different interface conditions: hot-on-hot (H/H), residual bitumen 200 g/m^2 (RB 200), and residual bitumen 400 g/m^2 (RB 400). The tests examined the bonding effects in terms of tangential force and shear displacement at failure, as well as the impact of vehicular traffic on rutting and fatigue failure. Finally, this study investigated the long-term aging effects of the binder on interlayer bonding and sought to correlate the results of numerical calculations with those of the laboratory tests.

Keywords: tack coat; bonding effect; bituminous interlayers; Leutner shear test; pavement layers

1. Introduction and Literature Review

From a global perspective, transportation infrastructure plays a fundamental role in the movement of people and goods, connecting centers and contributing to economic and social exchange. In particular, over the past years, roads and highways have witnessed continuous growth, driven by the ever-increasing demand for transportation and by the advent of new technologies and construction methodologies that led to the creation of safer, more functional, and durable infrastructures. In recent years, scientific research conducted in the field of road pavements has paid significant attention to the poorly understood topic of the adhesion between the layers of asphalt pavement [1–13]. For instance, in 2002, Mohammad et al. [1], evaluated the use of tack coats in asphalt pavements through controlled laboratory shear tests to determine the optimum application rate. Four emulsions (CRS 2P, SS-1, CSS-1, and SS-1 h) and two asphalt binders (PG 64-22 and PG 76-22 M) were selected as tack coat materials. The results showed that CRS-2P emulsion at 0.09 L/m^2 provided the maximum interface shear strength at both 25 °C and 55 °C. In 2009, another study [5] examined the long-term behavior of bonding properties in the lower layers of the pavement, finding that bond can decrease significantly over time. Results from the layer parallel shear test (LPDS) on Swiss roads show that rehabilitations or older constructions do not exhibit bond improvement, with shear force values potentially decreasing by up to 50% after 9–13 years.

Chun et al. [9], evaluated the impact of different interlayer bonding conditions on pavement performance and service life using finite element analysis and full-scale field tests, highlighting the importance of accurate modeling. Results showed that improved interlayer bonding positively affects pavement strain responses, enhancing structural performance and service life.

A study conducted on pavements in Malaysia [14] revealed that one of the major factors contributing to the degradation of the pavement is the poor bonding between layers, especially when it occurs between the wearing and binder courses, as compared to what happens along the deeper interfaces. Furthermore, degradation is amplified when high horizontal forces act in addition to vertical loads. This phenomenon can occur, for example, at road intersections, curves, slopes, and highway junctions where vehicles commonly brake and accelerate. Considering the simultaneous loss of pavement functionality due to fatigue and the accumulation of permanent deformation, the authors have found that the reduction in service life caused by poor bonding between wearing and binder courses is approximately from 94 to 98%.

A significant decrease in pavement service life may be also observed where bonding is poor between the base and binder layers courses. From a study conducted on the performance of pavements in Poland [15], it was observed that a low level of bonding between the base and binder courses leads to a simultaneous increase in tensile deformation (ε_t) on the intrados of these layers, resulting in a reduced fatigue resistance of the pavement and service life. From numerical calculations, it was observed that fatigue-induced failure initiates at the base of the binder layer, because the tensile deformations are greater than those observed at the bottom of the base layer. In addition, a low level of bonding also increases vertical deformations (εz) on the subgrade's extrados, increasing the likelihood of deep settlements leading to the subsequent surface rutting of the pavement. Hence, the need to design and build road pavements composed of well-connected layers to make a monolithic structure with an extended service life.

In 2014, the study on the behavior of bonded interfaces in bituminous pavements has highlighted interesting information regarding the physical and mechanical characterization of the pavement components [16]. In particular, it has been observed that the presence/absence of bonding coats, the type of bituminous emulsion, and the compaction method of the material play a fundamental role in the level of adhesion. The effect of each of these parameters on bonding has been evaluated in terms of tangential plain stress at failure following the Leutner shear test executed on laboratory-made samples as well as on cores extracted onsite. Specifically, the test results showed that choosing a bituminous emulsion composed of hard bitumen (50/70) provides a better connection between the wearing and binder courses compared to an equal amount of emulsion made out of soft bitumen (160/200). This result is justified by the fact that a bond coat made of soft bitumen generates a significant stiffness gradient between the bond coat itself and the layers of the pavement, which are made of a harder binder (35/50). The tangential stress value observed on specimens characterized by the presence of a soft residual bitumen at the interface is comparable to that one obtained on specimens without a bond coat. Furthermore, using a bond coat made of hard bitumen ensures better bonding than using none, except for the hot-on-hot laying technique. The latter promotes aggregate interlocking, thus improving the resistance to tangential shear. Under these conditions, in fact, the interlocking phenomenon at the interface becomes dominant compared to the tangential load causing the slipping of layers past each other. Often, the hot-on-hot technique allows the achievement of performances comparable to bond coats. In addition to the type and/or presence of the bond coat, a fundamental role in the connection between one layer to the other is played by the compaction method of the material. Laboratory tests have shown that compaction with a gyratory press guarantees the best bonding condition, as opposed to compaction of the same material with a static roller.

Laboratory compaction performed with a vibrating roller or Standard Compactor [17] leads to similar results. It has also been observed that the tangential resistance at the wearing and binder courses' interface improves with an increase in compaction level of the surface layer. Greater compaction of the wearing course, in fact, promotes the interlocking of the aggregates at the interface, thus leading to an increase in the shear strength required for the specimen's failure.

In the context outlined in the literature, the primary objective of the initial phase of this study is to assess, through a combination of numerical analyses and experimental investigations, the influence of the bonding between bituminous layers (in flexible road pavements) on their overall performance. More specifically, this research delves into the impact of bonding strength on the response to stresses induced by traffic, considering the continual growth in road transportation demands. The central focus of this investigation is on the bituminous layers of the pavement, where traffic-induced stresses are most pronounced and where maintaining exceptionally high performance is essential to avert impairments that could jeopardize road safety and driving comfort.

The research activity of the second part of this study refers to the materials, techniques, and regulatory requirements in force in Switzerland, where the analysis of bonding between pavement layers has gained importance in recent years and is the subject of study in many laboratories and research institutes. The Swiss standard VSS 40 430, 2022, which is used to verify the routine compliance of materials used in construction sites, currently prescribes a minimum value of tangential force at failure (measured at the interface between the layers through direct shear tests), which is indicative of the bonding strength [18]. However, the standard does not set a specific limit for the displacement at failure, permitting the material to reach the minimum tension regardless of the amount of relative slip between the layers. Through the investigations conducted, the impact of the level of bond between the bituminous layers on the durability of the superstructure is determined, as well as how the displacement at failure, which is obtained from mechanical characterization tests, controls the interface's performance.

2. Goals and Methodology

The scope of this study is divided into two parts: the theoretical investigation on the effect of tack coat on the structural behavior of a road pavement, followed by laboratory investigations on the performance of the tack coat in double layers (asphalt lab-made specimens and asphalt cores taken from site).

To accomplish the first goal, the theoretical behavior of a typical pavement section of Swiss highways has been determined using a mechanistic method. In addressing this topic, a rational calculation method tailored specifically for flexible pavements is adopted. This method focuses only on the layers' structural characteristics, making it a key reference tool for design and verification. The algorithm for the rational method is illustrated in Figure 1a.

The second goal is reached by conducting shear strength tests (Leutner shear test) on specimens both prepared in the laboratory and taken from an existing highway pavement. The Swiss standard procedures were adopted to characterize the mechanical behavior of the interface between the bituminous layers. Moreover, the mixtures used for preparing the specimens in the laboratory are those ones commonly specified for the construction of highway pavements in this country, which are designed to withstand significant and frequent loading cycles. The algorithm of laboratory investigations is presented in Figure 1b.

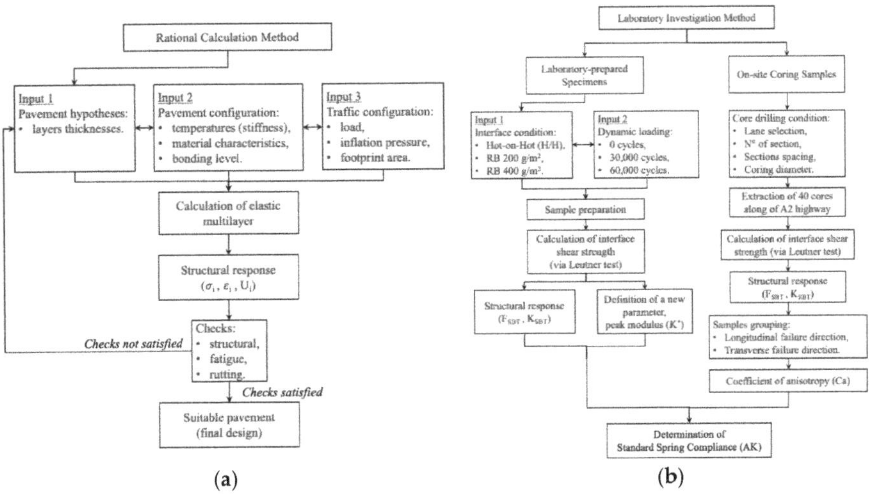

Figure 1. (**a**) Rational calculation method algorithm; (**b**) laboratory investigation algorithm.

3. Rational Calculation Method

To investigate the bonding between the double layers (wearing course and binder course), a numerical analysis was conducted. This analysis compared stress, deformations, and displacements at selected points within the pavement structure under varying thickness and mechanical properties [19]. To achieve this, a rational calculation method based on the theory of elastic multilayer systems was employed, consisting of two main steps.

First, the modeling of the pavement to calculate its structural response to vehicle loads and environmental conditions has been outlined. According to the Elastic Multilayer Theory, the pavement structure is represented as layers extending infinitely radially but finite vertically, except for the subgrade, which extends infinitely in both directions. Each layer is characterized by its thickness (t), stiffness modulus (E), and Poisson's ratio (ν).

Secondly, the prediction of the pavement's long-term performance was carried out using semi-empirical models to correlate the structural response with progressive deterioration.

While the study primarily focused on examining the interface between bituminous pavement layers, the calculated mechanical values highlighted the critical role that bonding between layers plays in determining the pavement structure's service life.

3.1. *Pavement Structure*

The modeled pavement has the typical layer stratigraphy of a Swiss highway pavement, as presented in Figure 2. It consists of four bonded asphalt courses and a granular mixed foundation layer (unbound), which is responsible for transmitting the stresses to the subgrade. Focusing on the bonded layers of the pavement, different types of mixtures are placed at different depths, each with specific characteristics and functions.

The Semi-Dense Asphalt (SDA) mixture is used for the wearing course, characterized by a noise-absorbing feature (to reduce tire–road noise). Additionally, modified bitumen of type PmB CH-E 45/80-65 is employed to further enhance the viscoelastic behavior of the mixture.

High modulus asphalt mixtures (EME, French: Enrobés à Module Elevé) are used for the base and the binder layers to provide high durability with heavy vehicular traffic, including trucks. The high stiffness mixtures reduce the pavement's thickness and maintenance costs [20]. The stiffness of these mixtures is achieved with graded "hard" binders (15/25 or 10/20) and selected (high performance, maximum packed) aggregates. This, combined with the correct binder dosage, results in a low void content (1% to 6%), which optimizes stress transmission. However, the use of very hard binders requires special attention in achieving optimal compaction, which is why they are only justified for pavements subjected to high dynamic loads. Specifically, high modulus bituminous mixtures, as specified by the SN 640 431-1-NA standard [21], serve the dual purpose of increasing fatigue resistance and limiting the accumulation of permanent deformations in the pavement.

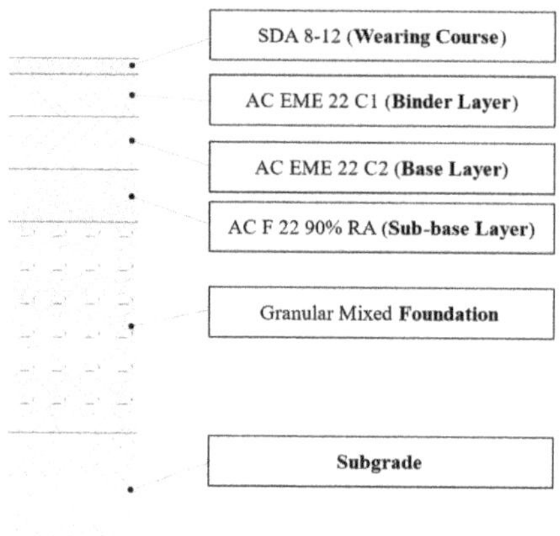

Figure 2. Stratigraphy of the modeled pavement.

Finally, the pavement structure ends with a fourth bitumen-bound layer that serves as the hot mix subbase. It is composed of the AC F 22 90% RA mixture, which is characterized by its composition of 90% recycled bituminous aggregate by weight.

3.2. Scenarios Definition

3.2.1. Layers Characteristics

For a thorough analysis of the model, various scenarios, each of which represents a specific geometric and mechanical condition of the pavement, are considered. These scenarios are defined according to the parameters listed in Table 1.

Table 1. Modeling parameters.

Layer Composition	Thickness Range According to VSS 40 430–40 436 [18] [mm]	Selected Thicknesses (t) [mm]	Stiffness Modulus (E) [MPa]			Poisson's Ratio (ν) [-]
			T [°C] = 10	T [°C] = 20	T [°C] = 30	
SDA 8–12 (Wearing Course)	25 to 40	30 and 40	4000	2500	700	0.35
AC EME 22 C1 (Binder Layer)	80 to 120	80, 90 and 100	14,000	8000	3000	0.35
AC EME 22 C2 (Base Layer)	80 to 120	100	17,000	12,000	6500	0.35
AC F 22 90% RA (Sub-base Layer)	60 to 150	100	15,000	10,000	5500	0.35
Granular Mixed Foundation	400	400	500	500	500	0.40
Subgrade	-	-	50	50	50	0.40

The thicknesses of the layers have been chosen in compliance with VSS 40 430–40 436 [18], which defines a reference range for each type of mixture. Based on this reference, thickness values are comparable with those commonly adopted in design practice, to keep the model as close as possible to real conditions. The 30 and 40 mm thick wearing courses and the 80, 90, and 100 mm thick binder layers are evaluated through this model. Since the stiffness moduli of the bituminous layers are dependent on temperature and loading frequency (at the same frequency, the higher the temperature, the lower the modulus) three different reference temperatures were selected.

The level of bonding between the bituminous layers is also considered as a further parameter to define the modeling scenarios, which can be defined by assigning a specific value to the Standard Spring Compliance (*AK*), with which the rational method models the behavior of the interface. This parameter is correlated to the horizontal reaction modulus (*K*) according to Equation (1).

$$AK = 1/K \tag{1}$$

In this equation, *K* expressed in N/m^3, or *K* [MPa/mm] = $1/(AK \times 10^{-9})$.

It is worth emphasizing that this parameter varies only at the interface between the wearing course and binder layer, while it remains constant for all the others, where an intermediate condition of Partial Bonding is always considered ($AK = 10^{-11}$ [m^3/N]). An exception is the Full Friction condition, which has been applied to all the interfaces of the road structure. Therefore, a total of three different bonding conditions, each of which corresponds to a specific value of the parameter *AK*, have been defined between the first two layers as presented in Table 2.

Table 2. Level of bonding at the wear–binder interface.

Level of Bonding (Wear–Binder Interface)	*AK* [m^3/N]
Full Friction	-
Partial Bonding	10^{-11}
Full Debonding	10^{-7}

It is important to specify that the Full Friction condition is purely ideal, where the layers are perfectly bonded to form a fully monolithic system. This eliminates relative displacements along the interface ensuring improved short-term and long-term performances. Also, it optimizes the overall mechanical behavior of the pavement, and, finally, it ensures extended durability. However, this ideal condition differs from what can actually be achieved in the field or in the laboratory.

The Partial Bonding condition ($AK = 10^{-11}$ m^3/N), although defined as "partial", is characterized by a horizontal reaction modulus (*K*) large enough to minimize displacements like for the Full Friction (complete bonding), but unlike the latter, it is actually reproducible in the lab or in the field.

Lastly, the extreme condition of Full Debonding ($AK = 10^{-7}$ m^3/N) is related to a scenario where the pavement's layers work independently of each other, resulting in significant relative displacements along the interface. In practice, this extreme condition never occurs thanks to the roughness of the surface of the layers in direct contact with each other providing always some residual shear strength of the interface.

3.2.2. Load Configurations

Two different load configurations have been modeled, reflecting the actual behavior of vehicle wheels traveling on the pavement, as presented in Figure 3.

Load configuration 1 involves the application of a vertical load (F_v) of 40 kN on a single wheel corresponding to an axial weight of 80 kN (standard axle load) (Figure 3a). Load configuration 2 involves the application of both a vertical load of 40 kN and a horizontal load (F_h) of 20 kN on a single wheel (Figure 3b). In this scenario, the horizontal force, mobilized in the direction of travel, models the accelerating force of vehicles (positive as per the selected reference system). This force arises from the bonding phenomenon at the interface between the vehicle wheel and the road pavement, influenced by the surface characteristics of the wearing course. Therefore, the horizontal force can be defined by Equation (2).

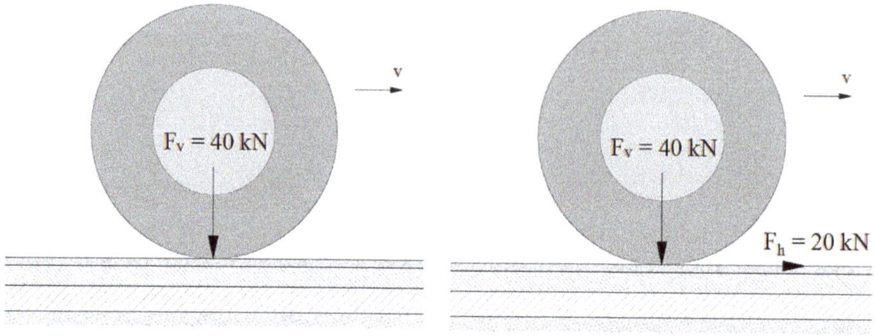

(a) (b)

Figure 3. (**a**) Load configuration 1; (**b**) load configuration 2.

$$F_h = f \cdot F_v \tag{2}$$

In this equation, a coefficient of friction for tire–pavement (f) equal to 0.5 [22] has been considered.

3.2.3. Pavement Stress–Strain Calculation Domain and Reference Points

The applied loads are distributed on the surface of the pavement over a circular contact area whose radius is inversely proportional to the tire inflation pressure. This aspect is crucial as loads distributed on the surface can lead to significant variations in the stress state within the pavement structure. To this end, a pressure of 0.8 MPa (8 bar) was applied, resulting in a contact area with a radius of 126 mm, as shown in Figure 4a.

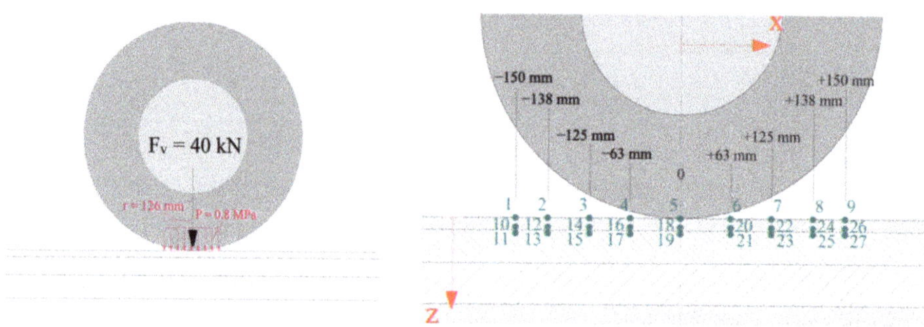

(a) (b)

Figure 4. (**a**) Load footprint; (**b**) reference points for calculating pavement mechanical properties.

The coordinates of the stress–strain calculation points refer to a Cartesian coordinate system {x, y, z}, where the x-axis runs along the direction of travel, the z-axis defines the depth from the surface, and the y-axis completes the right-handed coordinate system.

These reference points are distributed both on the pavement surface, and at the interface between the first two layers of the pavement, as shown in Figure 4b. This approach allows us to track the variation in stress and strain along the x-coordinate, making it possible to identify the maximum values. These maximum values were then selected as the primary parameters for comparison in the subsequent analyses.

3.3. Model Analysis

Considering the above-mentioned combinations of thickness and mechanical properties, 108 different scenarios were analyzed, and the results from the scenarios related to the

30 mm thick wearing course and the 80 mm thick binder course are reported below as examples. To carry out the simulation, as mentioned before, a rational calculation method based on the theory of elastic multilayer systems was employed, implemented using BISAR® 3.0 (Bitumen Structures Analysis in Roads), a software developed by Shell (Shell Global Solutions, based in The Hague, Netherlands) and widely utilized in professional practice for the design and verification of flexible and semi-rigid pavements. The simulation results are presented in Section 5.1.

4. Laboratory Investigation

The shear resistance between bitumen-bound pavement layers is evaluated using the Leutner shear test, as per the European standard EN 12697-48 [17]. Proposed in 1979 by Professor Rolf Leutner, this test is the primary method used in Germany and Switzerland to assess the bonding level between layers. It is preferred over other methods because it can test both laboratory-made specimens and field samples, and it is faster to perform. The Leutner test was utilized in the experimental phases of the current study.

The Leutner apparatus features two metal semi-rings: a fixed lower ring and a movable upper ring, which applies controlled displacement to the bituminous layer, as demonstrated in Figure 5. The test focuses on the shear behavior of the interface, which must be accurately positioned to reflect a condition of simple shear.

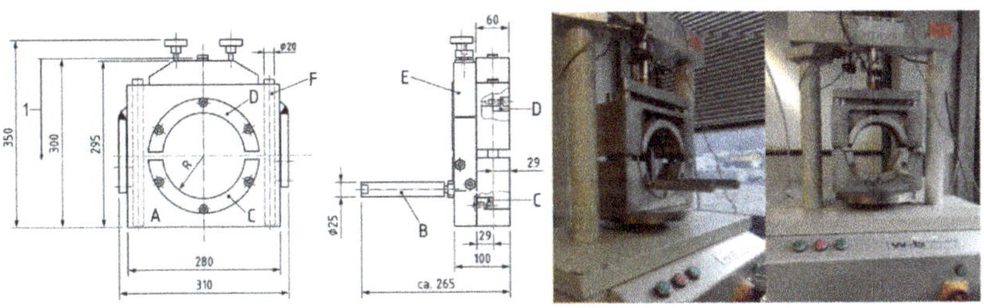

Figure 5. Leutner test equipment components [17].

In Figure 5, A is base body, B is sample support, C is lower shear ring, D is upper shear ring, E is upper body, F is guiding bar, and 1 shows a range max of ±20 mm. According to EN 12697-48 standard, the test typically uses cores with a diameter of 150 ± 2 mm and requires at least four samples for a representative series. The interface must be perpendicular to the core's axis, with a maximum deviation of 5 mm. The jaws holding the core must match or slightly exceed the core's radius by up to 2 mm.

Before testing, cores must be conditioned at 20 ± 1 °C for at least 4 h. The testing procedure involves assembling the Leutner frame, securing the core, and aligning the layer interface. The core is marked to indicate traffic or compaction direction. The press is started to record force and displacement data until the core breaks. The force–displacement curve is saved, and the layers are carefully separated post-test.

The standard mandates that the test must be completed within 15 min of removing the core from a temperature-controlled environment. The key measurements taken during the test include shear force, displacement, maximum shear force, the slope of the elastic linear segment, peak displacement, and the energies pre- and post-peak.

4.1. Laboratory-Prepared Specimen Experimentation

Specimens used in the experimentation were prepared in the laboratory by coring binder course's slabs. These slabs (500 × 180 mm module with a total height of 100 mm) were compacted using a Large Roller Compactor (Figure 6a). Each slab comprises a 40 mm thick surface layer (SDA 8–12) compacted over a 60 mm thick binder layer (AC EME 22 C1). This choice ensured continuity with the bituminous mixtures discussed in Section 3, maintaining consistency throughout the experimental phase. The study focused on analyzing three different interface conditions as the primary variables: no tack coat in Hot-on-Hot (H/H) conditions, tack coat with residual bitumen of 200 g/m^2 (RB 200), and tack coat with residual bitumen of 400 g/m^2 (RB 400). Additionally, three levels of dynamic loading were applied to the specimens: 0 cycles, 30,000 cycles, and 60,000 cycles.

(a) (b) (c) (d)

Figure 6. (**a**) Large Roller Compactor; (**b**) wheel tracker; (**c**) core drilling machine; (**d**) Specimen subjected to Leutner shear test.

The loading cycles were applied to the compacted slabs using a wheel tracker (Figure 6b) before coring the specimens. The program also included creating one slab for each combination of variables (n = 9). From each slab, 3 core samples with a diameter of 150 mm were extracted (Figure 6c), resulting in a total of 27 specimens. Each specimen was then subjected to shear testing using the Leutner shear test method (Figure 6d) to determine the shear strength of the interface under different imposed conditions. The results of laboratory-prepared specimens' experimentation are presented in Section 5.2.

4.2. Field Sample Experimentation

Alongside the investigation on laboratory-prepared specimens, the interface behavior between bituminous layers in an operational pavement was analyzed. Leutner shear tests were conducted on 40 cores extracted from a section of the A2 highway, a vital corridor crossing Switzerland from North to South and linking Southern and Northern Europe. This route experiences substantial daily traffic from heavy commercial vehicles, subjecting the pavement to significant stresses.

Tests were conducted on samples from an unloaded portion of the road section (emergency lane), assuming that the mechanical stresses on the examined material have been negligible in service. The coring of the 40 samples was concentrated along the emergency lane (South–North) near a rest area and its acceleration and deceleration lanes, between the Lugano Nord and Lugano Sud exits (chainage 24 + 400 km to 24 + 980 km), as depicted in Figure 7.

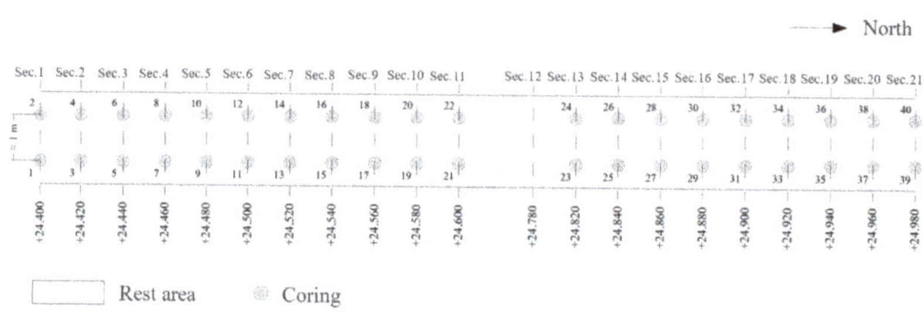

Figure 7. Scheme of core drilling performed along the A2 highway.

The cores are distributed along 20 pairs of sections, with each pair spaced 20 m apart. Within each section, the two cores are spaced approximately 1 m apart (Figure 8a). The coring process utilized a core drilling machine, which extracts samples using a rigid cylindrical steel motor-driven shaft. A core barrel at the end of the shaft penetrates the layers to obtain samples for the Leutner test (ϕ150 mm, variable length), as shown in Figure 8b.

(**a**) (**b**)

Figure 8. (**a**) Site inspection and coring points identification; (**b**) sample core drilling.

Data from a 2010 investigation identified the mixtures used for the surface and binder layers of a specific road section. The surface layer was constructed with an AC MR 11 mixture (macro-textured asphalt), characterized by a grain size distribution with a high percentage of aggregates between 8 and 11 mm in diameter, providing a highly frictional macro-texture suitable for a wearing course. The underlying binder layer used the HMT 22 mixture, now known as AC B 22.

The average thicknesses of the surface and binder layers were found to be 35 mm and 60 mm, respectively. The samples were divided into odd-numbered samples (longitudinal failure direction) and even-numbered samples (transverse failure direction) to identify an anisotropy coefficient linking failure values in the longitudinal and transverse directions. After conditioning at 20 °C, the samples underwent failure testing through the Leutner Shear Test. The results of field sample experimentation are presented in Section 5.3.

5. Results and Discussion

This section summarizes and discusses the results of both the modeling and the experimental phases, linking them through the common theme of bonding between bituminous layers and proposing new strategies to control performance. A reference value has been established for optimizing pavement mix designs based on the materials' mechanical characteristics. Additionally, using experimental data, a maximum displacement value at failure is defined. This, combined with the minimum tangential stress requirement from the VSS 40 430 standard (Leutner shear test), enhances the overall requirements for the shear behavior of the materials.

5.1. Outcomes of the Rational Calculation Method

5.1.1. Interface Shear Stress

Figure 9 represents the values of shear stress (τ_{xz}) generated at the wear–binder interface resulting from the rational calculation method, under the selected model parameters (different bonding conditions, loading configurations (1 and 2, as defined above), stiffness modulus, temperature, etc.).

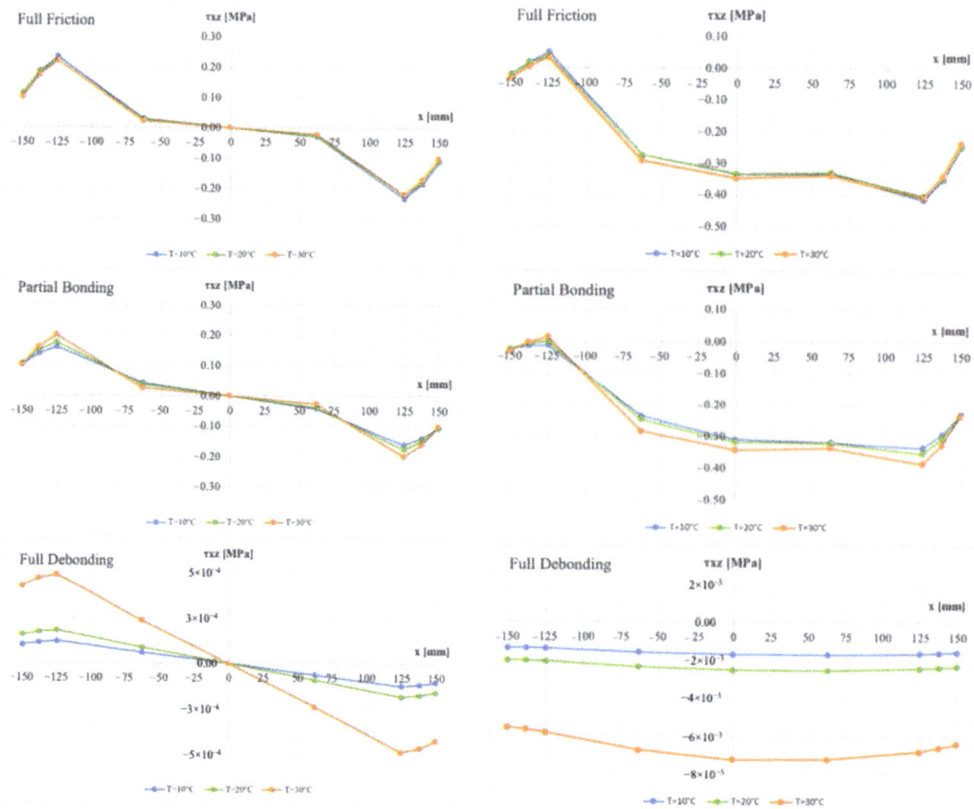

(**a**) (**b**)

Figure 9. Shear stress trend at the wear–binder interface, under different bonding conditions and temperatures (**a**) for loading configuration 1; (**b**) for loading configuration 2.

According to the results of loading configuration 1, it emerges that the shear stress τ_{xz} at the wear–binder interface reaches its maximum at 125 mm from the vertical axis of the wheel, both in the front and rear parts, as shown in Figure 9a. This is because the application of vertical loads only leads to the symmetry of stresses, which is consistent with the theory of elastic multilayers. The symmetry is with respect to the origin of the chosen reference system, for equilibrium and for the sign convention of the shear stresses.

Regarding loading configuration 2, the maximum shear stress is also located at 125 mm from the vertical axis of the wheel, but here, it is localized in the front part only. This is due to the introduction of a horizontal component of load on the surface, in addition to the vertical one, resulting in an asymmetrical stress distribution that varies depending on the intensity and direction of application of the shear force. Moreover, the abscissa of the maximum is 125 mm for both the conditions of Full Friction and Partial Bonding. In the condition of Full Debonding, this abscissa is 63 mm (half of the contact radius) as shown in Figure 9b.

It can be concluded that the asymmetry generated by the introduction of the horizontal component of load on the surface leads to a decrease in tension in the rear part of the wheel and to an increase in the front part, especially in the conditions of Full Friction and Partial Bonding. Thus, the maximum shear stress is always within the contact area and is localized near the edge, especially in the presence of a high bonding between the layers.

Therefore, it emerges that, for both loading configurations, the shear stress τ_{xz} takes on higher values as the bonding between the wearing and binder layers increases, that is, as the AK parameter becomes smaller (for the same temperature and thicknesses). In this regard, it is crucial to underline that the shear stress that occurs in the Full Debonding condition is three orders of magnitude smaller (two in the presence of horizontal load) compared to the stress that develops in the other two conditions with progressively higher levels of bonding.

5.1.2. Interface Horizontal Stress

Figure 10 represents the normal stress ($\sigma_{h,i}$) developed near the interface between the two bituminous layers as calculated from the rational method, and under the selected model parameters (different bonding conditions, loading configurations, stiffness modulus, temperature, etc.).

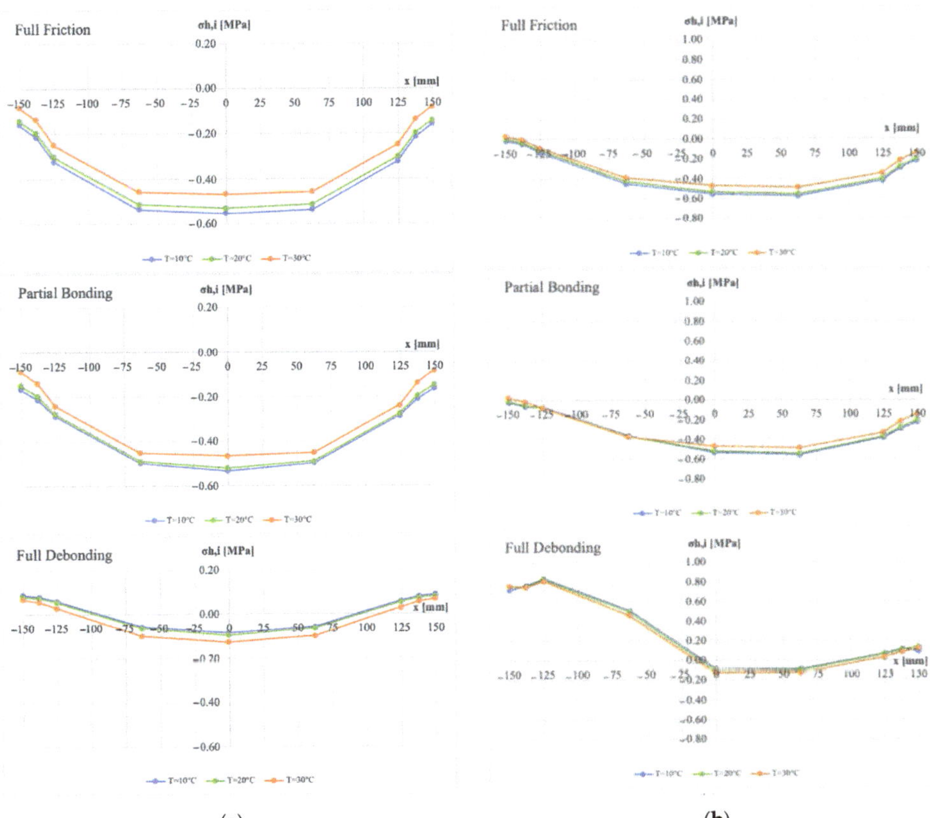

Figure 10. Horizontal stress trend at the wear–binder interface, under different bonding conditions and temperatures (**a**) for loading configuration 1; (**b**) for loading configuration 2.

The normal stress $\sigma_{h,i}$ (symmetric as well as the tangential stress) developed near the interface between the two bituminous layers reaches its maximum value at 150 mm from the wheel axis in those scenarios including the vertical load only. In scenarios involving both vertical and tangential loads, the absolute maximum value is reached at −150 mm (behind the wheel) in Full Friction and Partial Bonding conditions, and at −125 mm in the Full Debonding condition. This is true for all temperatures.

From the comparison between the two load configurations, it emerges that, in the presence of the horizontal force component, there is a significant variation in the normal stress ($\sigma_{h,i}$), which increases by an order of magnitude as compared to the vertical load only.

The results also show that in Full Debonding condition, tensile stresses arise at all temperatures and for both load configurations. In Full Friction and Partial Bonding, however, compression dominates at all temperatures when the pavement is subjected to the application of a vertical load of 40 kN only. While with the addition of the horizontal load, even if bonding is present, the stresses show positive values (tensile). In this regard, it is important to note that the increase in the value of the maximum horizontal stress refers not only to an increase in absolute tensile values but also to a decrease in compressive values. It is also evident that the Full Debonding condition is the most unfavorable since it maximizes, for the same load configuration, the value of the tensile stress at the interface, thus promoting cracking and the subsequent deterioration of the pavement.

From Table 3, it is possible to highlight several observations regarding the sensitivity of the interface horizontal stress to the variation of the thickness of the bituminous layers of pavement.

Table 3. Percentage variations of the horizontal stress at the interface.

Bonding Condition	**T [°C]**	**$t_{binder\ layer}$ = Variable, $t_{wearing\ course}$ = 40 mm D[$\sigma_{h,i}$] [%]**		**$t_{binder\ layer}$ = 100 mm, $t_{wearing\ course}$ = Variable D[$\sigma_{h,i}$] [%]**	
		Loading Config. 1	**Loading Config. 2**	**Loading Config. 1**	**Loading Config. 2**
Full Friction	10	2.74	10.40	−5.49	−477.28
	20	2.39	15.22	−6.33	−196.23
	30	1.49	25.78	−17.48	−55.37
Partial Bonding	10	1.99	14.29	−0.13	−342.49
	20	1.78	81.65	−2.12	−98.85
	30	1.20	14.10	−14.28	−47.53
Full Debonding	10	−1.91	−0.25	2.83	−32.33
	20	−1.34	−0.21	2.09	−32.76
	30	−0.25	−0.12	−5.23	−33.12

The mean percentage variation of stress with increasing thickness of the layers (D[$\sigma_{h,i}$]) is considered as an indicator of sensitivity, which can be computed through Equation (3).

$$D[\sigma_{h,i}](\%) = 100 \cdot \frac{\sigma_{mean}}{\sigma_{ref}}\left(t_{wearing\ course},\ t_{binder}\right) \quad (3)$$

It is evident from Table 3 that there is a greater sensitivity to a change in the thickness of the wearing course layer from 30 to 40 mm, particularly when both vertical and horizontal loads act simultaneously on the pavement. In this case, it can be observed that the stress at the interface decreases in all the modeled conditions of bonding and at all temperatures as the thickness of the wearing course increases. From Table 3, the high sensitivity of $\sigma_{h,i}$ with increasing thickness of the binder layer at a temperature of 20 °C needs to be recognized in the condition of Partial Bonding (+81.65%). In the condition of Full Debonding, however, the stress decreases at all temperatures and in both loading configurations.

5.1.3. Surface Horizontal Stress

The horizontal normal stress developed at surface ($\sigma_{h,s}$) reaches its maximum value in tension at 138 mm from the center of the imprint contact area symmetrically for loading configuration 1, and asymmetrically for configuration 2. In the latter, the maximum horizontal stress is localized 138 mm from the wheel axis in the rear for each bonding condition. Figure 11 compares the behavior of the maximum horizontal stress ($\sigma_{h,s}$) at the surface for the two considered loading configurations.

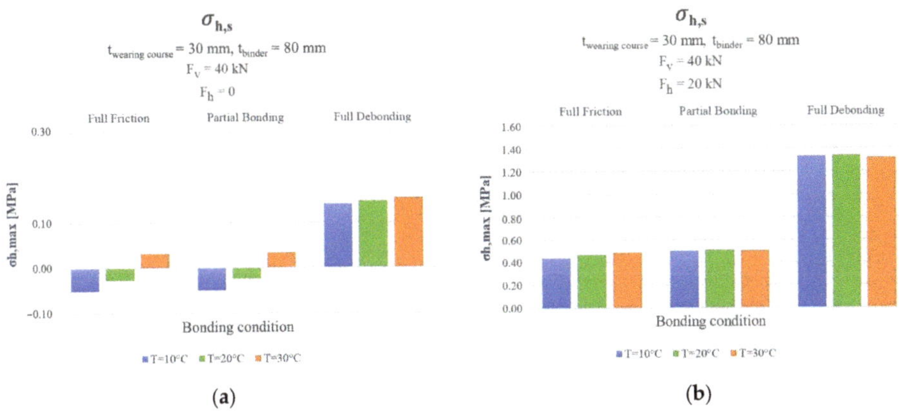

Figure 11. Horizontal stress at surface, under different bonding conditions and temperatures (**a**) for loading configuration 1; (**b**) for loading configuration 2.

From the graphs, it can be observed that the surface horizontal stress reaches the highest values in the Full Debonding condition, in both loading configurations. Moreover, it is evident that the stress values, for a given bonding condition, are greater when there is a horizontal force at the surface. In fact, as far as loading configuration 2 is concerned, the resulting stresses are tensile at the interface for every bonding condition and at all temperatures. On the other hand, with the first loading configuration both in Full Friction and Partial Bonding, the horizontal stress is tensile only at 30 °C, while at other temperatures, tension is never reached. This is different in Full Debonding, where tensile stresses are present at every temperature.

It is noteworthy that, when subjected to the vertical load only, the stress magnitude rises with temperature, irrespective of the bonding condition. However, when a horizontal force is present, the stress increases with temperature only in the Full Friction condition. In the other two cases, the stress reaches the peak value at 20 °C. Being this maximum value negligible, it is possible therefore to consider stress as constant with temperature under such conditions.

5.1.4. Interface Relative Sliding

The behavior of the relative sliding at the interface (ΔU_X) is evaluated at the point of maximum shear stress, where the sliding itself is also maximum (using the horizontal reaction modulus K). Firstly, it is important to highlight that this quantity is defined by Equation (4).

$$\Delta U_x = U_x(wearing\ course) - U_x(binder) \tag{4}$$

In this equation, ΔU_X represents the magnitude of displacement at the interface of the upper layer (wearing course) relative to the underlying one (binder layer).

Figure 12 represents the values obtained for the relative sliding at the interface, under the adopted model parameters (different bonding conditions, loading configurations, stiffness modulus, temperature, etc.).

According to Figure 12, the displacement is higher as the bonding level decreases, at the same temperature. Indeed, in Full Friction, there is no relative sliding (ΔU_X).

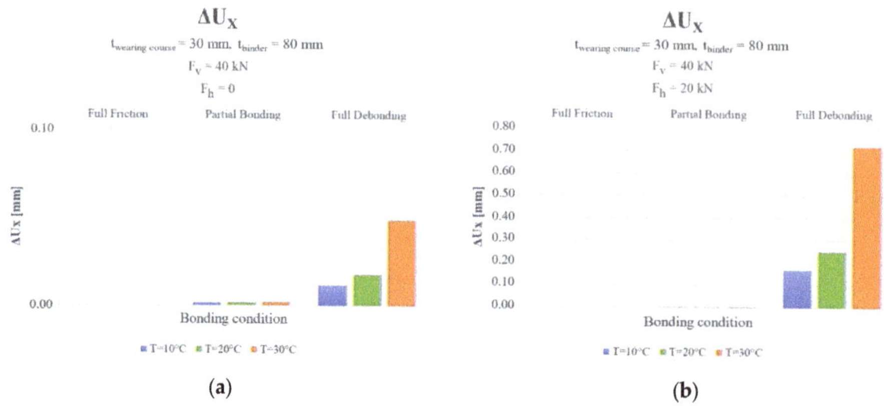

Figure 12. Relative sliding at the interface during (**a**) loading configuration 1; (**b**) loading configuration 2.

Moreover, for the same bonding level, temperature, and layer geometry, larger relative sliding is generated by the horizontal force of 20 kN, in agreement with the increase in the shear stress τ_{xz} for this loading configuration. According to the results, it can be noted that:

- For the vertical load only, the values obtained in Full Debonding are one order of magnitude higher than those obtained in Partial Bonding;
- For the horizontal force in Partial Bonding, the values remain of the same order of magnitude as those obtained with the first loading configuration;
- With Full Debonding, on the other hand, relative displacement is 10 times larger compared to that obtained from the application of the vertical load only.

Therefore, it is confirmed that the presence of a horizontal load, such as that caused by the braking or the accelerating forces of traveling vehicles, can significantly impact the relative slippage between the layers, especially when the bonding at the interface is on the lower end of spectrum.

5.2. Results of Experimentation on Laboratory-Prepared Specimens

Figure 13 summarizes the results obtained from the Shear Bonding Tests (SBT), under three different load cycles and interface condition scenarios, in terms of maximum breaking force ($F_{SBT,max}$) and horizontal reaction modulus (K_{SBT}), which corresponds to the slope of elastic linear portion of the failure curve.

It was observed that the maximum breaking force (shear breaking force according to the Leutner test) increases with the load cycles under the same interface condition, indicating an improved bonding strength between the surface and underlying layers over time. This increase is likely due to post-compaction from vehicle traffic. However, after this period, the breaking force values of the cores show a decline over time.

The necessity of a tack coat when laying down bitumen pavement layers can be influenced by temperature. In hot-over-hot paving conditions, the bituminous materials are more fluid and have lower viscosity, which can potentially improve bonding between layers without a tack coat. However, this does not mean that a tack coat is unnecessary. Even in hot conditions, a tack coat can ensure a stronger and more reliable bond, minimizing the risk of layer separation under traffic loads. According to Graziani et al. [23], the interlayer shear stress (ISS) values are significantly improved with a tack coat at lower paving temperatures (5 °C), highlighting its importance in ensuring proper layer adhesion, although the improvement is less significant at higher paving temperatures (40 °C). In fact, in low paving temperatures, bituminous materials become stiffer and less adhesive, increasing the risk of poor bonding between layers. In such conditions, a tack coat is crucial to compensate for the reduced bonding capacity.

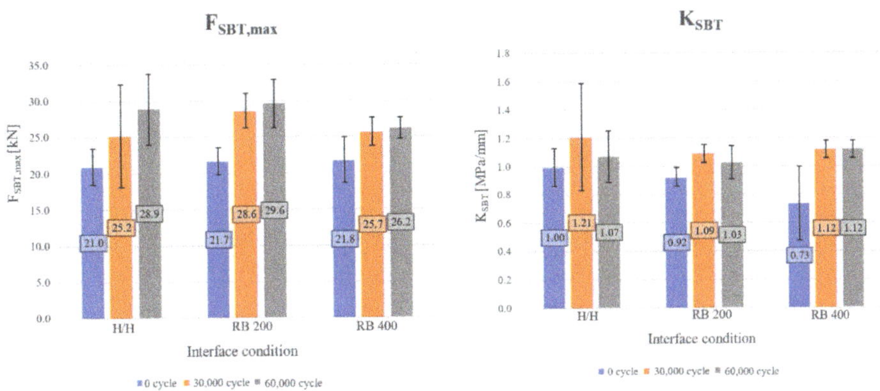

(**a**) (**b**)

Figure 13. Obtained results from specimens subjected to the Leutner shear test: (**a**) average values of breaking force; (**b**) horizontal reaction modulus.

This trend suggests that the breaking force tends to increase with t load cycles initially, without a sudden decrease. This result is plausible given the differences between laboratory testing conditions and actual site conditions. Laboratory conditions cannot simulate the long-term aging of the binder caused by environmental exposure. Over time, this aging in actual pavements could lead to a gradual decrease in the interface's resisting force. Additionally, a high number of vehicular loads, combined with aging, can cause rutting and fatigue failure, promoting delamination between pavement layers and thus reducing shear resistance.

The test results indicate that a residual bitumen content of 200 g/m^2 (RB 200) is optimal for bonding. Under the same load cycles, this condition has achieved the highest value of the breaking force. An exception is noted for specimens not subjected to load cycles, where an interlayer content of 400 g/m^2 yields a slightly higher shear force. However, the differences in breaking forces between these conditions are negligible.

For specimens subjected to cyclic traffic loading, the horizontal reaction modulus varied significantly, ranging from 0.45 MPa/mm to 1.72 MPa/mm. The mean values, as shown in Figure 13b, remain relatively constant to around 1 MPa/mm, with an overall mean value of 1.03 MPa/mm. This corresponds to a bonding parameter AK value of approximately 10^{-9} m^3/N. As discussed in Section 3, this AK value indicates a Partial Bonding condition between the layers.

Although the K_{SBT} parameter indicates the bonding level between layers by defining the slope of the linear portion of the failure curve and the AK parameter used in rational calculations, it does not provide information about the peak force and displacement values from experimental tests. Mechanistic methods are based on simplifying assumptions that approximate the behavior of pavement materials to be elastic linear.

However, laboratory mechanical tests reveal the elastoplastic nature of bituminous mixtures and their interfaces. To further analyze failure behavior, the peak force and displacement values of the F-δ curve were considered to derive a new modulus value (K*), representing the average behavior of the interface at failure. This parameter is the slope of the line passing through the origin and tangent to the failure curve at its maximum point, as shown in Figure 14a.

Figure 14b presents a scatter plot of the peak force and displacement values from tested specimens under three different interface conditions. A linear regression was performed on these data, with the intercept at the origin, yielding a modulus K* value of 0.65 MPa/mm.

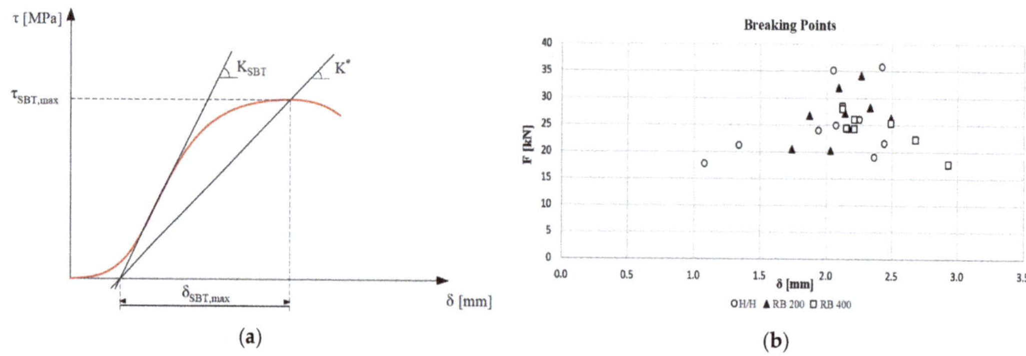

Figure 14. (**a**) K_{SBT} and K* modules; (**b**) breaking points obtained from the Leutner test.

Figure 14b reveals significant dispersion in results due to the substantial variability in the K_{SBT} modulus and displacement, underscoring the variable nature of the K* modulus. This dispersion likely stems from the limited number of specimens and the suboptimal reliability of the chosen compaction method. Compaction using a Roller Compactor often struggles with achieving volumetric homogeneity in the produced slab, which, along with the manual brushing method used for spreading the interface material, could have significantly impacted the results' variability.

Despite this dispersion, the K* value obtained from linear regression is considered a global reference parameter. Comparing the elastic modulus (K_{SBT}) with the peak modulus (K*), a slight deviation is observed, confirming that bituminous materials exhibit some ductility, where the mechanical behavior of the interface becomes plastic during failure.

5.3. Results of Experimentation on Field Samples

Figure 15 presents the average values of the test results conducted on all 40 core samples, in terms of maximum breaking force ($F_{SBT,max}$) and horizontal reaction modulus (K_{SBT}).

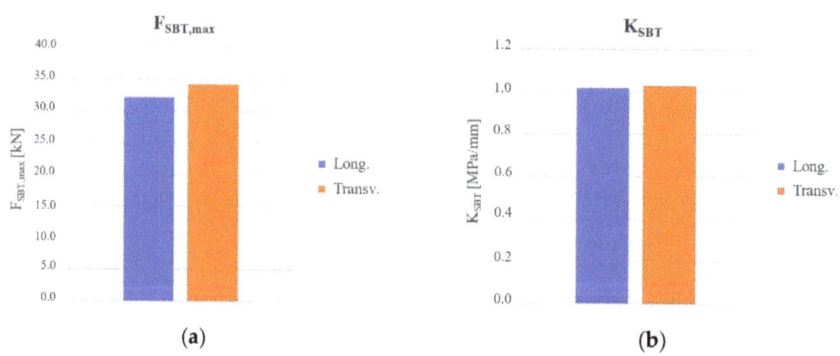

Figure 15. Obtained results from samples subjected to the Leutner shear test: (**a**) average values of breaking force; (**b**) horizontal reaction modulus.

As seen in Figure 15, there are no significant differences observed between the values along the two different failure directions. In terms of peak force ($F_{SBT,max}$), values ranging from 32 kN (longitudinal direction) to 34 kN (transverse direction) were recorded, resulting in only a 6% difference between the two. Likewise, the value of horizontal reaction modulus (K_{SBT}), has a negligible 1% variation observed between the longitudinal and transverse failure conditions. The parameter stabilizes at a value of 1 MPa/mm, corresponding to a value of AK equal to 10^{-9} m^3/N. Thus, once again, it is evident that the AK parameter takes a value comparable to that one found in the laboratory-prepared specimens.

Thanks to the distinct group of results, it is possible to relate the average values obtained to define a coefficient of anisotropy (C_a) that connects the data recorded along the two different cutting directions. To encompass the magnitude of the breaking force and the displacement, the average values of the accumulated energy during the test before reaching the peak condition (E_{SBT}) were used. In particular, the value of the coefficient C_a is defined by Equation (5).

$$C_a = \frac{E_L}{E_T} \tag{5}$$

In this equation, E_L is the average pre-peak energy along the longitudinal direction; and E_T is the average pre-peak energy along the transverse direction.

Figure 16 shows that the accumulated energy before failure is similar in both directions: 60.95 Nm in the longitudinal direction and 64.20 Nm in the transverse direction, resulting in an anisotropy coefficient (C_a) of 0.95. This suggests that the failure behavior of the interface is not significantly affected by the direction of the tangential force. The C_a value does not indicate the influence of vehicular traffic, as the pavement section sampled was not subject to vehicle loads. However, it provides insight into the method of laying the surface layer over the underlying one. The near-unity anisotropy coefficient indicates that the surface and binder layers were laid in succession with a tack coat of bituminous emulsion, confirming that there was no milling of the wearing course and its reapplication. This is supported by the fracture surface characteristics of the lower layer, which lacks the longitudinal grooves typical of milled surfaces. According to the authors' experience, a milled surface would show higher pre-peak energy values (or breaking force) in the transverse direction due to increased interlocking from the grooves, enhancing the overall shear strength of the interface.

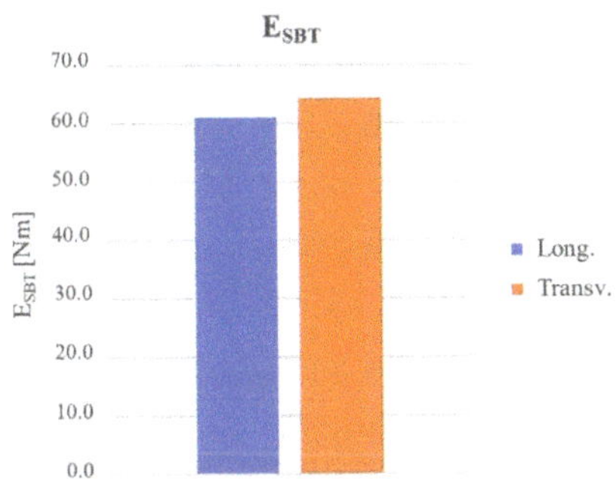

Figure 16. Average energy at failure for samples subjected to Leutner shear test.

5.4. Comparison between Modeling and Experimental Investigation

The numerical analysis discussed in Section 3, using a rational calculation method, provided valuable insights into the characterization of the connection between the bituminous layers of pavement. It was observed that the choice of bonding condition significantly influences the overall stress–strain state of the superstructure, highlighting the need to test the bond strength of bituminous materials for informed design decisions. The analysis revealed that the maximum tangential stress at the interface at 20 °C under the most severe condition is 0.40 MPa, which is significantly lower than the 1.43 MPa observed in laboratory-prepared specimens and below the Swiss standard VSS 40 430 requirement of 0.85 MPa.

However, the numerical modeling and experimental investigation do not align closely enough for a direct comparison of calculated and measured quantities. The numerical model assumes a constant load and an elastic linear, homogeneous, isotropic material, whereas the Leutner test involves a gradually applied tangential force until failure, where the material exhibits plastic behavior.

Despite these differences, the calculation software remains a valid tool for simulating the behavior of layered superstructures like road pavements. The Leutner test, while conservative, provides shear strength in the "plane shear" mode like in situ shear strength tests on soils. Although it cannot fully simulate the operational behavior of a pavement, the modeling of the superstructure helps in understanding the experimental results.

The AK parameter (or K_{SBT}) obtained from tests allows, finally, for the accurate and unequivocal identification and classification of the bonding level between layers.

5.5. Design Reference and Limit Displacement

While tack coats positively impact shear resistance, they are insufficient for achieving optimal bonding conditions. Modeling results indicate that the best bonding conditions enhance the superstructure's durability. However, the experimental analysis suggests using the AK parameter obtained from specimens as a reference value in road design. This approach avoids overestimating pavement service life by assuming overly optimistic AK values or designing under (unrealistic) Full Friction conditions. Overestimation could lead to early repairs under budget constraints. Therefore, it is here recommended to use an AK value of 10^{-9} m^3/N between the wearing and binder layers during the design phase to ensure the superstructure modeling closely aligns with the materials' actual performance.

Once the AK parameter design value was determined, focus shifted to meeting the minimum requirements specified by Swiss standards for the shear performance of bituminous mixtures. The standard mandates a minimum tangential stress of 0.85 MPa at failure, which corresponds to a shear force of 15 kN on the cylindrical specimens of ϕ150 mm tested in the Leutner test. Notably, the standard does not specify requirements for failure displacement or other control parameters like the elastic modulus K_{SBT}. Consequently, two key values were compared for reference:

- The minimum tangential stress (τ_{min}) equals to 0.85 MPa;
- The peak modulus (K*) equals to 0.65 MPa/mm.

The comparison focused on K* and τ_{min} because both are critical to determining the interface failure condition. Although the peak modulus is an independent parameter, it strongly correlates with the previously chosen design parameter. Specifically, in this case, the AK value of 10^{-9} m^3/N (corresponding to K_{SBT} = 1 MPa/mm) aligns with a K* value of 0.65 MPa/mm. Following the investigation, it is anticipated and required that Leutner tests conducted on a significant number of samples to control shear performance will yield average failure values aligning along the slope of the K* peak derived from linear regression of experimental results. Combining this requirement with the minimum failure stress defined in Figure 17 establishes a specific limit displacement value (δ_{lim}). This value is determined by the intersection of the peak line (K*) and the horizontal line $\tau = \tau_{min}$ = 0.85 MPa, resulting in δ_{lim} equal to 1.30 mm in this case.

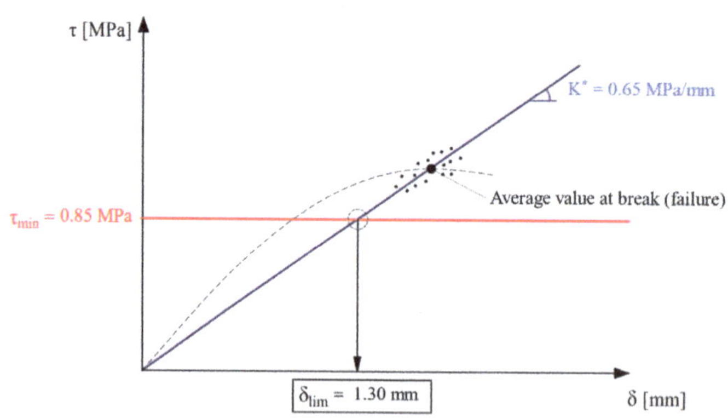

Figure 17. Definition of the limit displacement at failure.

This value has dual significance. Firstly, it represents the maximum displacement value required by the failure test to meet both imposed requirements. Specifically, with K* set as a design reference, if $\delta_{SBT,max}$ is less than δ_{lim}, it indicates that the failure stress is below the minimum required. Secondly, δ_{lim} is the maximum displacement achievable under the minimum tangential stress. Exceeding this displacement for the same force would result in a K* value smaller than that determined by the investigation, thereby negatively impacting the material's operational performance.

6. Conclusions

This study combined numerical modeling and experimental investigation to explore the bonding at the interface of bituminous layers in road pavements. Using a rational calculation method, the research analyzed how the road superstructure response varies with design parameters (stiffness, layer thickness, load configuration) and bonding conditions. Modeling the response of a Swiss highway with high traffic, this study demonstrated that the interlayer connection (AK parameter) and external load configuration significantly influence pavement performance. Poor bonding leads to increased displacements and tensile forces, negatively impacting durability. The study concludes that accurate pavement design must consider interlayer bonding.

The three different conditions (Full Friction, Partial Bonding, and Full Debonding) provide a comprehensive understanding of the extremes within which the actual performance of the pavement layers can vary. By exploring these boundary scenarios, the study can identify the worst-case and best-case performance outcomes and provide a range of expected behavior that helps in risk assessment and management. These extreme conditions serve as a foundation for future research by identifying key parameters that significantly influence pavement performance. Further studies can then focus on optimizing these parameters for better outcomes.

BISAR® (Bitumen Structures Analysis in Roads), which is used for these simulations, is a widely recognized and validated software tool in the field of pavement engineering, frequently used by researchers and practitioners for its robust and reliable calculations. The material definitions and system assumptions within BISAR are based on extensive research and have been thoroughly validated in numerous studies (e.g., [24,25]).

The experimental investigation identified an optimal quantity of tack coat that enhances the connection between bituminous layers, as determined by the tangential force at failure from the Leutner test. Additionally, cyclic loading was found to improve the interlayer bond strength. The analysis of the horizontal reaction modulus (K) indicated that, regardless of the interface condition (tack coat presence/absence) and the number of loading cycles, the interface always exhibits Partial Bonding. This has led to defining a reference value of AK as 10^{-9} m^3/N for design purposes. By examining the material's failure behavior, a limit displacement value at peak condition was established, complementing the existing requirement of the minimum tangential force at failure for a more comprehensive design parameter.

Additionally, sampling from an inactive section of the Swiss A2 highway (i.e., emergency lane) revealed that the bonding condition remains partial, even in an environmentally aged and unloaded pavement. Furthermore, in this specific case, the direction of the tangential force causing specimen failure did not affect the Leutner test results.

Author Contributions: Conceptualization, E.T. and M.K.; methodology, V.G.G. and G.M.; validation, E.T., V.G.G. and M.K.; formal analysis, A.S. and G.M.; investigation, A.S. and C.R.; resources, M.D.T.; data curation, A.S., C.R. and M.K.; writing—original draft preparation, A.S.; writing—review and editing, E.T., V.G.G. and M.K.; supervision, E.T. and M.D.T.; project administration, E.T. and M.D.T. All authors have read and agreed to the published version of the manuscript.

Funding: This research received no external funding.

Data Availability Statement: Data are contained within the article.

Conflicts of Interest: The authors declare no conflicts of interest.

References

1. Mohammad, L.N.; Raqib, M.; Huang, B. Influence of asphalt tack coat materials on interface shear strength. *Transp. Res. Rec.* **2002**, *1789*, 56–65. [CrossRef]
2. Sangiorgi, C.; Collop, A.C.; Thom, N.H. Laboratory assessment of bond condition using the Leutner shear test. In Proceedings of the 3rd International Conference on Bituminous Mixtures and Pavements, Thessalonkiki, Greece, 21 November 2002; pp. 315–324.
3. Bognacki, C.J.; Frisvold, A.; Bennert, T. Investigation of asphalt pavement slippage failures on runway 4R-22L, Newark International Airport. In Proceedings of the 2007 Worldwide Airport Technology Transfer Conference, Atlantic City, NJ, USA, 16–18 April 2007; Federal Aviation Administration American Association of Airport Executives: Washington, DC, USA, 2007.
4. Muench, S.T.; Moomaw, T. *De-Bonding of Hot Mix Asphalt Pavements in Washington State: An Initial Investigation*; TransNow UTC: Berkeley, CA, USA, 2008.
5. Raab, C.; Partl, M.N. Interlayer bonding of binder, base and subbase layers of asphalt pavements: Long-term performance. *Constr. Build. Mater.* **2009**, *23*, 2926–2931. [CrossRef]
6. Plug, C.P.; Bondt, A.H.; Hydra, R. Optimum bonding of thin asphalt overlays on porous asphalt. In Proceedings of the 5th World Congress on Emulsions, Lyon, France, 12–14 October 2010.
7. Vaitkus, A.; Žilionienė, D.; Paulauskaitė, S.; Tuminienė, F.; Žiliūtė, L. Research and assessment of asphalt layers bonding. *Balt. J. Road Bridge Eng.* **2011**, *6*, 210–218. [CrossRef]
8. Panda, M.; Sutradhar, B.; Giri, J.; Chattaraj, U. An experimental study on assessment of pavement interlayer bond strength. *Int. J. Transp. Sci. Technol.* **2013**, *2*, 141–147. [CrossRef]
9. Chun, S.; Kim, K.; Greene, J.; Choubane, B. Evaluation of interlayer bonding condition on structural response characteristics of asphalt pavement using finite element analysis and full-scale field tests. *Constr. Build. Mater.* **2015**, *96*, 307–318. [CrossRef]
10. Wang, X.; Su, Z.; Xu, A.; Zhou, A.; Zhang, H. Shear fatigue between asphalt pavement layers and its application in design. *Constr. Build. Mater.* **2017**, *135*, 297–305. [CrossRef]
11. Wan, J.; Wu, S.; Xiao, Y.; Amirkhanian, S.; Liu, G.; Hu, J.; Zhang, D. Using interface shear characteristics to evaluate the residual performance of asphalt pavements. *J. Adhes. Sci. Technol.* **2018**, *32*, 2070–2082. [CrossRef]
12. Rahman, A.; Huang, H.; Ai, C. Effect of shear stresses on asphalt pavement performance and evaluating debonding potential under repeated loading: Numerical study. *J. Transp. Eng. Part B Pavements* **2021**, *147*, 04021023. [CrossRef]
13. Yu, J.; Cheng, G.; Zheng, Y.; Zeng, T.; Hu, X. Experimental Study on Interface Model of Asphalt Pavement under Vertical Load. *Appl. Sci.* **2022**, *12*, 8579. [CrossRef]
14. Zulkifili, N.H.; Sutanto, M.H. The Influence of Bonding between Layers on Pavement Performance, a Case Study of Malaysian Road. *E3S Web Conf.* **2018**, *65*, 09002. [CrossRef]
15. Jaskula, P.; Rys, D. Effect of interlayer bonding quality of asphalt layers on pavement performance. In *IOP Conference Series: Materials Science and Engineering*; IOP Publishing: Bristol, UK, 2017; Volume 236, p. 012005.
16. Jaskuła, P. Influence of compaction effectiveness on interlayer bonding of asphalt layers. In Proceedings of the 9th International Conference "Environmental Engineering", Vilnius, Lithuania, 22–24 May 2014.
17. *SN EN 12697-48*; Mélanges Bitumineux—Méthodes D'essai—Partie 48: Collage de Couches. BSI Standards Limited: London, UK, 2022.
18. *VSS 40 430*; Enrobés Bitumineux Compactés. Swiss Association of Road and Traffic Professionals VSS: Zurich, Switzerland, 2022.
19. Li, S.; Tang, L.; Yao, K. Comparison of two typical professional programs for mechanical analysis of interlayer bonding of asphalt pavement structure. *Adv. Mater. Sci. Eng.* **2020**, *2020*, 5850627. [CrossRef]
20. Celauro, C.; Ranieri, M.; Bonventre, F. Conglomerati ad alto modulo (eme) migliorati con additivi plastici. *Mecc. Mater. Strutt.* **2014**, *4*, 31–45.
21. *SN EN 13108-1*; Mélanges Bitumineux—Spécications des Matériaux—Partie 1: Enrobés Bitumineux. 2022.
22. Lushinga, N.; Xin, J. Effect of horizontal shear load on pavement performance. In Proceedings of the 2nd International Conference on Geological and Civil Engineering, Dubai, United Arab Emirates, 10–11 January 2015; pp. 13–14.
23. Graziani, A.; Canestrari, F.; Cardone, F.; Ferrotti, G. Time–temperature superposition principle for interlayer shear strength of bituminous pavements. *Road Mater. Pavement Des.* **2017**, *18* (Suppl. 2), 12–25. [CrossRef]
24. Al-Qadi, I.L.; Carpenter, S.H.; Leng, Z.; Ozer, H.; Trepanier, J. *Tack Coat Optimization for HMA Overlays: Laboratory Testing*; FHWA-ICT-08-023; Illinois Center for Transportation: Rantoul, IL, USA, 2008.
25. West, R.C.; Zhang, J.; Moore, J. *Evaluation of Bond Strength between Pavement Layers (No. NCAT Report 05-08)*; Auburn University, National Center for Asphalt Technology: Auburn, AL, USA, 2005.

buildings

MDPI

Article

Research on the Embedding Behavior of Ceramic Particles on the Surface of Epoxy Resin Anti-Skid Thin Layer of Pavement

Jiaquan Yuan [1], Gang Zhou [2,*], Chaoliang Fu [3,*], Wenhong Duan [1], Yifan Zhang [2], Haoyang Huang [2], Weihong Jiang [1], Li Xiong [1], Huimei Li [1], Xiaohua Yang [1] and Chuanqiang Li [4,*]

1 Dali Danan Highway Co., Ltd., Dali 671000, China; yuanjiaquan2024@163.com (J.Y.); duanwenhong0703@163.com (W.D.); jiangweihong_cn@163.com (W.J.); xiongli_dali@163.com (L.X.); lihuimei_cn@163.com (H.L.); yangxiaohua_cn@163.com (X.Y.)

2 School of Civil Engineering, Chongqing Jiaotong University, Chongqing 400074, China; zhangyifan8691@163.com (Y.Z.); hhy@mails.cqjtu.edu.cn (H.H.)

3 Institute of Highway Engineering, RWTH Aachen University, 52074 Aachen, Germany

4 School of Materials Science and Engineering, Chongqing Jiaotong University, Chongqing 400074, China

* Correspondence: cjhs_2000@163.com (G.Z.); fu@isac.rwth-aachen.de (C.F.); lichuanqiang@cqjtu.edu.cn (C.L.)

Abstract: Colored polymer anti-skid thin layers are widely used on urban roads to enhance driving safety, improve road aesthetics, and mitigate the urban heat island effect. However, in thin layers constructed by the spreading method, the adhesion of cementitious material to the aggregate is often weak. This leads to early-stage spalling of surface aggregates, thereby reducing the anti-skid performance of the layer. To investigate the factors contributing to spalling, this study examines the embedding behavior of ceramic particles and assesses how the fluidity of the cementitious material and aggregate shape characteristics influence the embedding depth. Using a rotational viscosity test, it is concluded that a cementitious mix ratio of adhesive/powder filler/sand filler = 1:0.5:1 or 1:0.5:1.5 facilitates effective aggregate embedding. Testing the embedding depth of aggregates with the same particle size across different cementitious materials revealed that higher cementitious viscosity results in a reduced aggregate embedding depth. Geometric parameter data for aggregate particles were extensively collected using an image acquisition device, and quantitative analysis identified the shape characteristics influencing the embedding depth. A gray correlation analysis determined that the impact of the shape characteristics on embedding depth follows the order of roundness factor > prism factor > axial coefficient.

Keywords: ceramic particles; epoxy resin; embedding depth; image processing; colored anti-skid thin layer

1. Introduction

Roads are composed of multiple layers, each with specific functions, among which the anti-skid performance of the pavement is a crucial factor affecting the safety of vehicle driving [1]. However, as the surface layer of the pavement is the outermost layer that directly comes into contact with vehicles and the natural environment, under the long-term effects of ultraviolet rays, oxygen, water, and vehicle loads, the pavement structure will be damaged, mainly manifesting as deformation, cracking, spalling, and surface polishing, which results in a decrease in the anti-skid performance of the pavement, significantly affecting the driving safety of vehicles and shortening the service life of the road [2–5]. Therefore, properly repairing existing pavement distresses, preventing their continuous deterioration, and enhancing the anti-skid performance of pavements have become key problems that need to be urgently solved in highway maintenance and management. The ultra-thin wearing layer, as a preventive maintenance measure, can quickly repair pavement damage and enhance the anti-skid performance of pavements and has been widely used in highway engineering construction [6–8]. Ultra-thin wearing layers are typically constructed by blending various binders and aggregates, with modified asphalt and resin-based polymers frequently used as the primary binding agents. Compared to ordinary asphalt, epoxy resin-based binders have excellent bonding performance, good durability, good compatibility with various pavement materials, and better road performance, effectively extending the service life of the pavement coating [9]. The colored ultra-thin wearing layer is a form of colored pavement and is laid using polymer resin materials, such as epoxy resin and acrylic, as binders and adding appropriate colored ceramic particles [10]. The colored ultra-thin wearing layer not only provides protection for the pavement but also features improved anti-skid properties and enhanced visibility, both of which reduce accident rates. Additionally, it helps mitigate the urban heat island effect and enhances the aesthetic appeal of the city [11,12].

Currently, surface coating, color mortar, and spreading methods are mainly used for laying colored polymer anti-skid thin layers on highway pavements. Among the above-mentioned common construction methods, the surface coating method has achieved industrialization with a higher degree of mechanization, and the technical solutions for construction applications are relatively complete and mature. However, the colored mortar and spreading methods rely more on manual operation. Although this method has stronger mobility and higher adaptability, it still needs to be standardized and unified in terms of the construction process and material index [13]. Some studies have shown that a thin anti-skid layer laid by the spreading method has the best anti-skid performance [14,15].

Throughout long-term service, the colored ultra-thin wearing layer may experience issues such as aggregate loosening, color fading, reduced anti-skid performance, rutting, and interface bonding failure, all of which can diminish the layer's effectiveness in providing skid resistance [16]. Studies have demonstrated that the surface structure of the road is an important factor affecting the anti-skid performance of the pavement [17,18]. From a macroscopic perspective, to accurately describe the surface structure of the road, the geometric statistical indicators usually used by researchers are the mean profile depth (MPD) and mean texture depth (MTD), which have good synergy with the BPN and can characterize the anti-skid performance of the pavement to a certain extent [19–21]. However, the surface structure of the pavement is affected by many factors, such as the maximum size of the aggregate, shape of the aggregate, asphalt content, gradation, and thickness of the asphalt film [22]. Using the MPD and MTD cannot completely reflect the surface structure and anti-skid performance of the pavement. From a microscopic perspective, some scholars have studied the influence of the aggregate shape on the pavement structure. It was found that the prism of fine aggregates will affect the anti-skid performance of the asphalt mixture [23,24].

At the same time, for the colored ultra-thin wearing layer constructed by the spreading method, aggregate spalling is the main manifestation of thin layer damage. Kim et al. [25] proposed a model that could predict and evaluate the stripping process in the mixture. Boulangéet al. [26] analyzed the interface stripping mechanism between asphalt and aggregates using physicochemical methods and proposed that increasing the proportion of aromatic components and colloids in asphalt is beneficial to the adhesion of the asphalt–aggregate interface, thereby reducing the water sensitivity of the mixture. Most studies on the stripping mechanism are based on adhesion or stripping tests, focusing on the stripping of asphalt aggregates by water [27], while research on the embedded state of aggregates is lacking. For a given aggregate and binder, the degree to which the aggregate is embedded in the binder determines the behavior of the aggregate–binder interaction [28]. Kumbargeri et al. [29] used finite element analysis to study the influence of the aggregate shape and embedding percentage on the performance of the chip seal. Ozdemir et al. [30] quantified the degree of aggregate embedding using image processing methods. Seitllari and Kutay [31] developed a prediction model for changes in the embedding depth of a chip seal using multiple linear regression. In addition, the embedding depth determines the surface tectonic depth and spalling property of the ultra-thin layer. The deeper the embedding depth of the aggregate on the binder surface, the better the anti-spalling property. But, the surface tectonic depth of the layer will be decreased, leading to a decrease in anti-skid performance. However, if the embedding depth is shallow, the aggregates are prone to peel off, which may also be harmful to the anti-skid performance of the thin layer [32]. Therefore, the embedding depth plays a key role in the anti-skid performance of the ultra-thin wearing layer on the pavement.

Given that, this study investigates the factors affecting the embedding depth of aggregates in epoxy resin, focusing on the viscosity of the thin layer binder and the shape characteristics of the aggregates. The embedding depth of the aggregates within the thin layer is analyzed, and the influence of the binder viscosity and aggregate shape characteristics is assessed. Using image processing software, the shape characteristics of different aggregates with identical particle sizes are quantified. Finally, the degree to which each shape characteristic index impacts the embedding depth is examined. The flow chart of this work is shown in Figure 1.

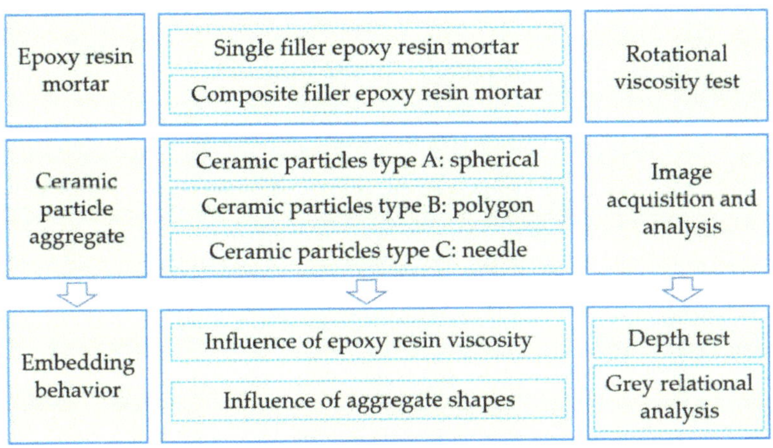

Figure 1. Flow chart of the research process.

2. Experiments

2.1. Materials and Specimen Preparation

The materials were water-based composite modified resin, inorganic powder filler (mainly composed of components such as carbonate and sulfate, with a particle size of >120 mesh), colored sand sintered from quartz sand and pigment, and colored ceramic particles (Figure 2a). The colored sand was treated using an 80–120 mesh sieve to ensure its uniform particle size.

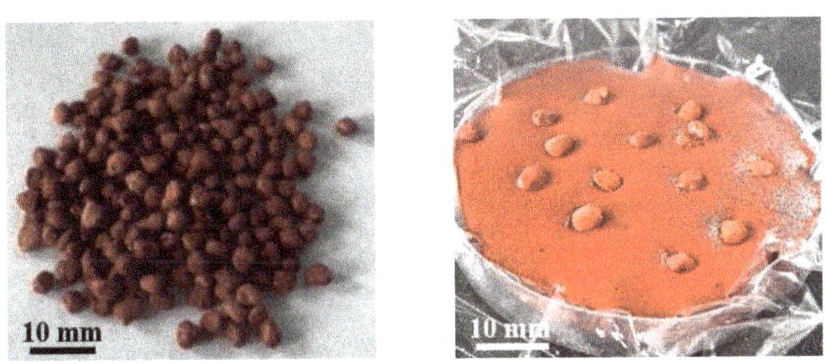

(**a**) Ceramic particles (**b**) Specimens

Figure 2. Ceramic particles and specimens.

The specimen was shown in Figure 2b, and they were prepared as follows: First, mix the powder, sand, and resin adhesive in a specified ratio for 5 min to form a slurry. Then, pour the slurry into a circular mold with a diameter of 50 mm and a height of 10 mm. Based on the target spreading height, freely distribute the ceramic particles into the mold from a height of 70 cm. Finally, cure the specimens indoors at 25 °C for 3 days.

2.2. Rotational Viscosity Test

The rotational viscosity test was carried out with reference to "Plastics-Epoxy Resins-Determination of Viscosity" (GB/T 22314-2008) [33]. The working principle is that the motor drives the rotor to rotate at a constant speed. The torque measurement device can measure the torque required for the rotor to overcome the viscous resistance of the sample during rotation. The viscosities of the tested samples were obtained after calculation and processing using the sensor.

2.3. Embedding Depth Measurement Test

Some scholars have measured the embedding depth, perimeter, and area of the aggregates in the chip seal by cutting the specimens and using section image processing to represent the embedding rate [32]. Although the colored anti-skid thin layer is constructed using the spreading method, which is similar to the construction process of the chip seal, the aggregates spread in the colored anti-skid thin layer are mainly single small-sized 2–5 mm ceramic particles. After the specimen is cut, it will cause serious damage to the embedding state of the aggregates, which makes image processing difficult. Therefore, to measure the embedding depth of the aggregates constructed by the spreading method, a test method for the embedding depth of the aggregates based on the thickness measurement was designed. The specific test method is as follows:

① As shown in Figure 3, on the surface of the cured specimen, color the exposed part of the aggregates in the thin layer to distinguish the critical edge where the aggregates contact the thin layer.

② Break the thin layer to take out the aggregates and polish the part of the aggregates that are not stained with sandpaper to obtain the aggregates to be tested.

③ Place the stained part of the aggregates facing down and press it into an ESP high-density foam board with a size of 300 × 100 × 20 mm until the stained part is completely pressed in. The foam board can fix the state of the aggregates after embedding.

④ After the aggregates are pressed into the foam board, the exposed part that is not stained on the surface of the foam template is the part where the aggregates are embedded in the thin layer. Use a thickness gauge to measure the total thickness of the aggregates and foam board after they are embedded and fixed in the foam board. Subtract the thickness of the foam board from the total thickness to obtain the embedding depth of the aggregates in the thin layer.

2.4. Aggregate Contour Information Processing Method Based on U-Net Network Model

Ceramic particles of 3 mm in size were classified into three types, A, B, and C, according to their different shape characteristics. Aggregate images were collected using a self-made image acquisition device, as shown in Figure 4. Using a previously trained U-Net network model, two-dimensional contour image information of the aggregates was obtained. Based on the contour information image, numerous geometric parameters of the aggregate particles were obtained, and the shape characteristics of the aggregate particles were quantified using geometric parameters.

First, the aggregate image dataset completed by manual annotation was deeply trained in the U-Net network model. The U-Net network code was written in the Python language. The network structure was built with the help of the neural network TensorFlow library and Keras library, and image processing was carried out using the OpenCV (version 4.1) and Scikit-image libraries (version 0.22.0).

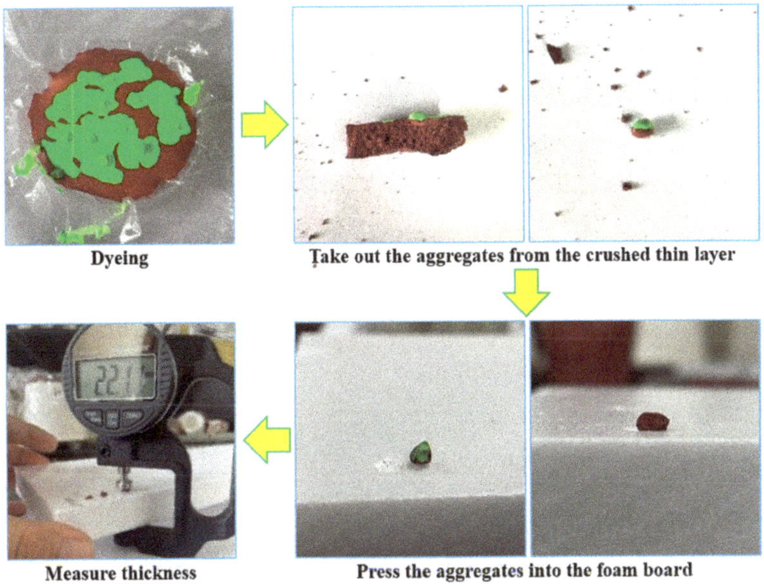

Figure 3. Measurement steps for aggregate embedding depth.

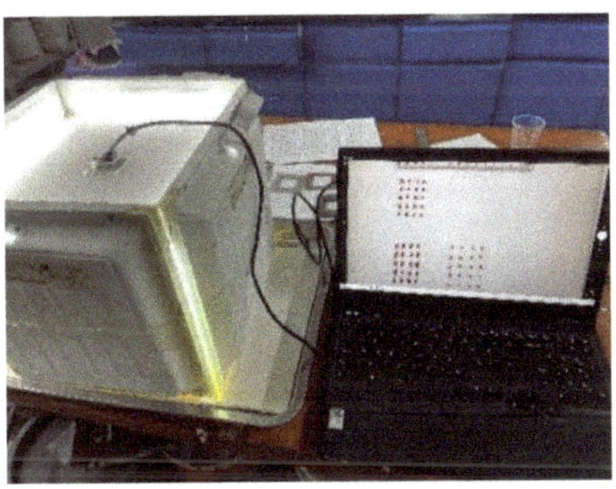

Figure 4. Image acquisition device.

After obtaining the aggregate contour map, geometric parameter information was extracted using the OpenCV image processing code library. Because the unit of the relevant parameter information of the aggregate contour calculated by the function in the OpenCV code library is pixels, to establish the connection between pixels and length units, the length of the ruler was photographed using an image acquisition device, and the number of pixels corresponding to 1 mm of the ruler was measured using ImageJ image processing software (version 1.54g) to calculate the conversion relationship between pixels and length units. As shown in Figure 5, the ImageJ measurement results show that a 1 mm length in the captured image corresponds to 14.5 pixels. Figure 6 shows the aggregate contour maps before and after processing.

According to the processed aggregate contour map, the geometric parameter information was extracted using the OpenCV image processing code library, and the axial coefficient, roundness factor, and prism factor of each aggregate were calculated in batches. The mean values of the shape characteristic parameters of the three types of aggregates were calculated.

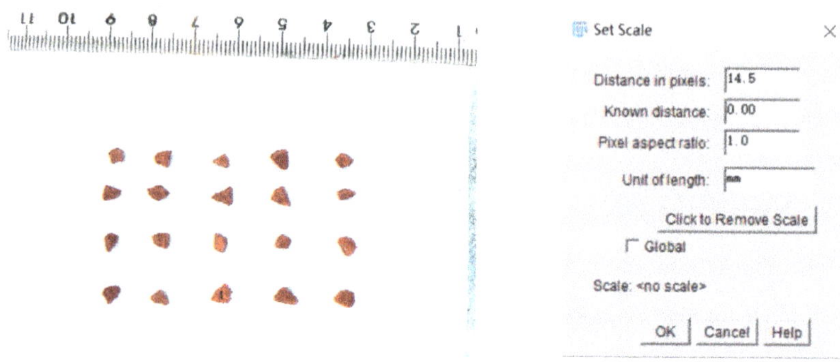

(**a**) Particles (**b**) Dimension value

Figure 5. ImageJ pixel point measurement.

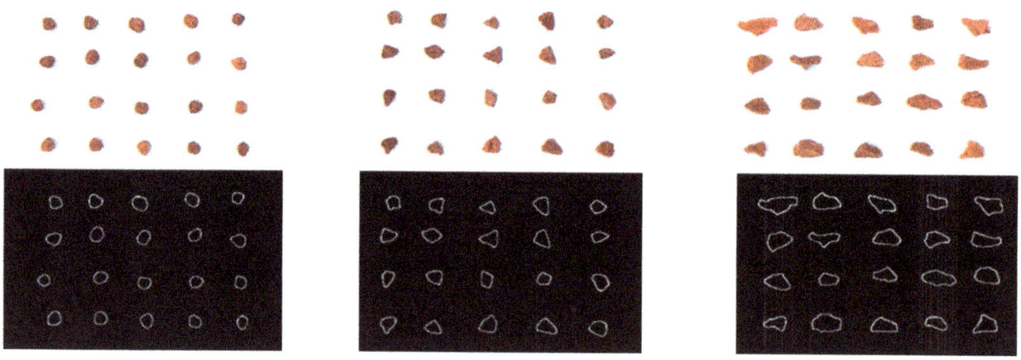

(**a**) Type A (**b**) Type B (**c**) Type C

Figure 6. Three types of aggregate contour maps.

The formulas for calculating the axial coefficient, roundness factor, and prism factor are as follows:

$$AR = \frac{l}{w} \quad (1)$$

$$RC = \sqrt{\frac{4\pi A}{P^2}} \quad (2)$$

$$PF = \frac{4\pi A}{P^2} \quad (3)$$

where AR represents the axial coefficient, RC is used to denote the roundness factor, PF refers to the prism factor. The symbol "l" indicates the length of the circumscribed rectangle of the aggregate, "w" represents the width of the circumscribed rectangle of the aggregate, "A" stands for the average area of the aggregate (see Table 1), and finally, "P" is used to represent the perimeter of the aggregate.

Table 1. Average area and corresponding standard deviations of each aggregate group.

	Type A	Type B	Type C
Average area	7.07	8.61	15.24
Standard deviation	0.96	1.73	3.57

2.5. Gray Relational Analysis

The shape characteristic index of the aggregate was taken as the characteristic sequence, and the embedding depth of the three types of aggregates in the thin layer was taken as the mother sequence. The degree of influence of the three shape characteristic indexes of the aggregate on the embedding depth of the aggregate was determined using a gray relational analysis. The specific steps of the gray relational analysis are as follows:

① Determine the characteristic sequence and mother sequence

Characteristic sequence (various factors affecting the evaluation criteria):

$$[X_1\ X_2\ \cdots\ X_n] = \begin{bmatrix} x_1(1) & x_2(1) & \cdots & x_n(1) \\ x_1(2) & x_2(2) & \cdots & x_n(2) \\ \vdots & \vdots & & \vdots \\ x_1(m) & x_2(m) & \cdots & x_n(m) \end{bmatrix} \tag{4}$$

Mother sequence (evaluation criteria):

$$X_0 = (x_0(1), x_0(2), \cdots, x_0(m))^T \tag{5}$$

② Unification of dimensions

Because different indicators usually have different orders of magnitude and measurement units, to eliminate the influence of these differences, it is necessary to perform dimensionless standardization processing on the original data, convert them into dimensionless values, and set all indicator data at the same order of magnitude to ensure the consistency and comparability of the analysis.

③ Calculate the correlation coefficient

Calculate the correlation coefficient of each factor according to the following formula γ:

$$\Delta\min = \min_i \min_k |x_0(k) - x_i(k)| \tag{6}$$

$$\Delta\max = \max_i \max_k |x_0(k) - x_i(k)| \tag{7}$$

$$\Delta_{ik} = |x_0(k) - x_i(k)| \tag{8}$$

$$\gamma(x_0(k), x_i(k)) = \frac{\Delta\min + \rho\Delta\max}{\Delta_{ik} + \rho\Delta\max} \tag{9}$$

where ρ denotes the resolution coefficient. The smaller its value, the greater the resolution. Typically, it is taken as $\rho = 0.5$.

④ Calculate the degree of correlation

Calculate the weighted average values of the correlation coefficients of the corresponding elements of the reference sequence:

$$\gamma_{0i} = \frac{1}{m}\sum_{k=1}^{m} \gamma(x_0(k), x_i(k)) \tag{10}$$

⑤ Analyze the data

Compare the gray-weighted correlation degrees and establish the correlation sequence between the evaluation and the object.

3. Results and Discussion

3.1. Influence of the Proportion of Thin-Layer Cementing Materials on Viscosity

3.1.1. Influence of Single Filler on the Viscosity of the Thin Layer

The influences of the powder and sand dosages on the viscosity of the thin layer were analyzed separately. The measured viscosity, selected rotor, and rotational speed are shown in Figure 7, where the filler dosage (%) is the dosage relative to the resin adhesive.

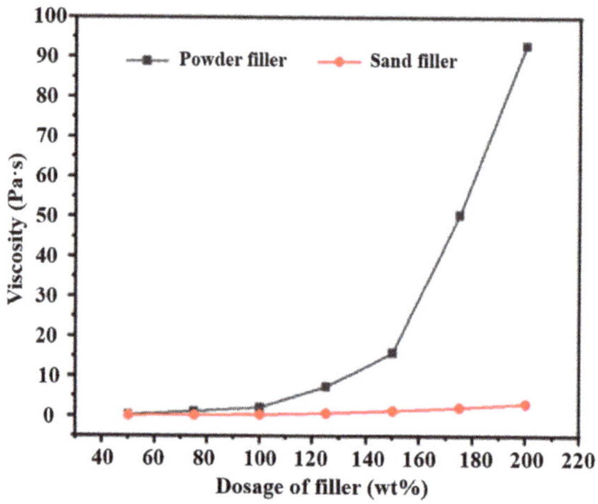

Figure 7. Effect of single filler on viscosity of thin layers.

When either powder or sand filler was added individually, the viscosity of the thin-layer cementitious material increased with higher filler dosages, indicating that the filler dosage had a direct effect on material viscosity. When the powder filler is incorporated at a 50% dosage, the viscosity of the cementitious material measured 0.41 Pa·s. As the dosage increased to 75%, 100%, 125%, 150%, 175%, and 200%, the viscosity rose to 1.83 Pa·s, 7.43 Pa·s, 15.59 Pa·s, 50.19 Pa·s, and 92.69 Pa·s, respectively. Beyond a 100% dosage, the growth rate of the viscosity accelerated, suggesting that internal voids in the material reached saturation, causing further powder additions to sharply increase the viscosity. In comparison, when the sand filler is added at the same 50% dosage, the viscosity of the cementitious material is 0.2 Pa·s. With increased dosages to 75%, 100%, 125%, 150%, 175%, and 200%, the viscosity rises at a slower rate, reaching increases of 0.12 Pa·s, 0.28 Pa·s, 0.66 Pa·s, 1.3 Pa·s, 2.08 Pa·s, and 3.06 Pa·s, respectively. This slower rate of viscosity increase, compared to powder filler, indicates that the sand filler has a smaller impact on the viscosity of the adhesive. At equivalent dosages, the powder filler resulted in a notably higher viscosity growth rate than the sand filler, highlighting that the powder filler can more effectively increase the viscosity of cementitious material. Specifically, at dosages of 50%, 75%, 100%, 125%, 150%, 175%, and 200%, the viscosity of powder-filled material exceeded that of sand-filled material by 0.21 Pa·s, 0.93 Pa·s, 1.76 Pa·s, 6.62 Pa·s, 14.5 Pa·s, 48.32 Pa·s, and 89.84 Pa·s, respectively. This gap in viscosity between powder and sand fillers continues to grow as the dosage increases. In summary, at the same dosage, the powder filler has a more substantial effect on increasing the viscosity of cementitious material than the sand filler. The smaller particle size of the powder filler allows it to fill voids more effectively, enhancing material compactness. Additionally, the powder filler exhibits slight chemical interactions with the adhesive, and its smaller particles (such as $CaCO_3$) increase the winding density of resin macromolecular chains. In contrast, the larger sand particles do not effectively fill voids, only physically blend with the adhesive, and are prone to sedimentation and segregation, resulting in a minimal effect on viscosity.

3.1.2. Influence of Composite Fillers on the Viscosity of the Thin Layer

The powder filler enhances the adhesive's mechanical properties through chemical interaction, while the sand filler provides a dense structure and boosts anti-skid performance. However, excess powder can cause agglomeration, and too much sand may lead to stratification, both affecting uniformity and performance. Adding a small amount of powder filler alone can effectively improve the performance of the adhesive; however, a large amount of resin is required, which increases the engineering cost. The addition of sand filler alone has an insignificant effect on the performance of the adhesive, and if the dosage is too large, it will cause a decrease in the performance of the cementitious material. Therefore, it is necessary to use a combination of powder and sand fillers to improve the performance of cementitious materials and control engineering costs. When the dosage of the powder filler exceeds 100%, it has a greater influence on the viscosity of the cementitious material, and the upper limit of the dosage of the powder filler is controlled at 100%. The effects of the composite fillers on the viscosities of the cementitious materials are listed in Table 2.

Table 2. Rotational viscosity with composite filler doping.

Ratio (Adhesive: Powder: Sand)	Rotational Viscosity (Pa·s)	Rotor#	Speed (r/min)
1:0.5:0.5	0.88	27	50
1:0.5:1	1.73	27	50
1:0.5:1.5	3.11	27	50
1:0.5:2	9.64	27	20
1:1:0.5	3.28	27	50
1:1:1	8.82	27	20
1:1:1.5	18.80	27	10
1:1:2	45.00	27	5

The effect of composite filler on the viscosity of thin layers is shown in Figure 8. When the powder filler content was maintained at 50%, increasing the sand filler content initially led to a gradual rise in the viscosity of the cementitious material. However, when the sand filler content exceeded 150%, the rate of viscosity increase accelerated sharply. Specifically, at sand filler contents of 100%, 150%, and 200%, the viscosities increased by 0.85, 2.23, and 8.76 Pa·s, respectively, compared to the viscosity at a 50% sand filler content. With a powder filler content of 100%, the internal voids in the cementitious material approached saturation, resulting in a more pronounced increase in viscosity as the sand filler content was raised. In this case, the sand filler contents of 100%, 150%, and 200% led to viscosity increases of 5.54, 15.52, and 41.72 Pa·s, respectively, relative to a 50% sand filler content. Comparing the effects of 50% and 100% powder filler contents, both the overall viscosity and the rate of viscosity increase were greater with a higher powder filler content, with the difference in viscosity between the two scenarios widening progressively as the sand filler content increased. In conclusion, with a baseline of 50% powder filler, a sand filler content above 150% results in a rapid increase in viscosity. To achieve more controlled adjustments of cementitious material viscosity when combining powder and sand fillers, it is advisable to use a lower powder filler content while regulating the viscosity through the sand filler content. This approach enhances control over viscosity changes and prevents large variations resulting from minor filler adjustments.

3.2. Influence of Thin-Layer Viscosity on the Embedment Depth of Aggregates

The viscosity of the thin-layer cementitious material will affect the construction performance; simultaneously, when using the spreading method for construction, it will affect the embedment depth of the spread aggregates. If the embedment depth of the aggregates is relatively shallow, the anti-stripping ability of the aggregates is mainly provided by the thin layer's bonding, which lacks encapsulation of the aggregates. Under these conditions, the anti-stripping ability of the aggregates is poor, and they are prone to stripping when subjected to loads. Therefore, when mixing the thin-layer cementitious material, it is necessary to appropriately adjust the viscosity of the cementitious material. If the viscosity is too high, the aggregates cannot be embedded in the thin layer and stripping is likely to occur. If the viscosity is too low, the thin layer will be too thick, and the aggregates will be completely buried in the thin layer and unable to play an anti-skidding role.

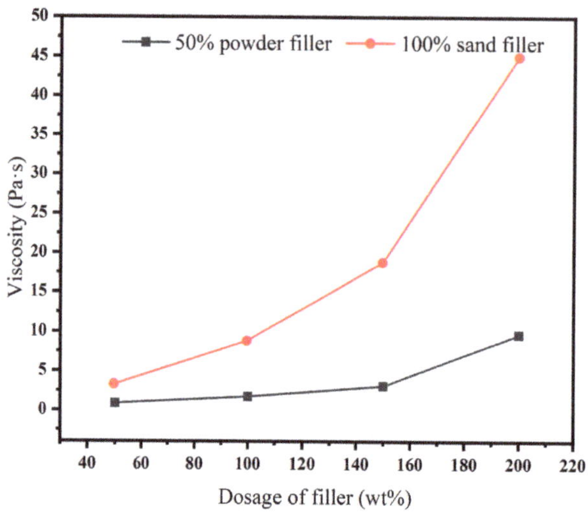

Figure 8. Effect of composite filler on the viscosity of thin layers.

The eight groups of composite filler ratios listed in Table 2 were used to mix and obtain thin-layer cementitious materials with different viscosities. The embedment states of the aggregates in thin layers with different ratios of adhesive/powder filler/sand filler is shown in Figure 9. To minimize the influence of the aggregate shape characteristics on the embedment depth, colored ceramic particles with a particle size of 3 mm and a relatively round shape were uniformly selected for spreading.

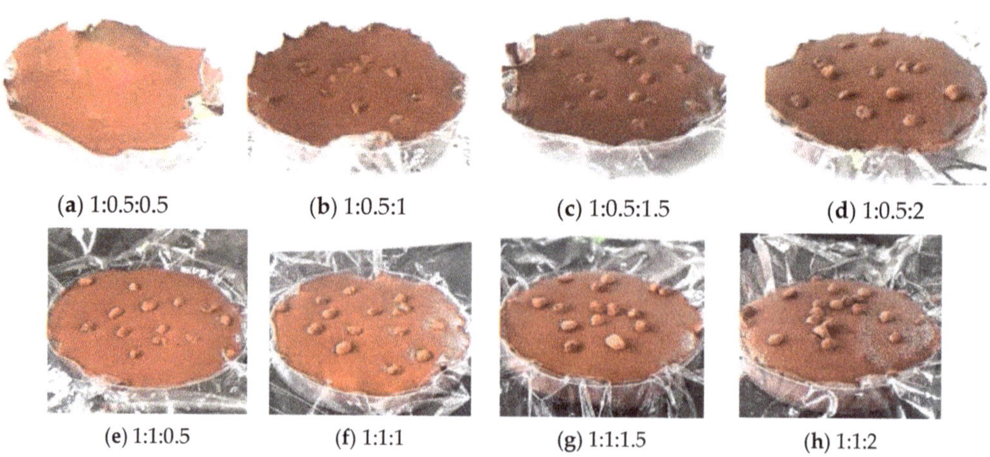

Figure 9. Embedding state of aggregates in thin layers with different ratios.

After spreading the aggregates, 10 aggregates were collected from each group, and their embedding depth was measured. When the aggregates were completely embedded in the thin layer, the embedding depth was taken as 3 mm. The embedded depths of the aggregates are shown in Figure 10.

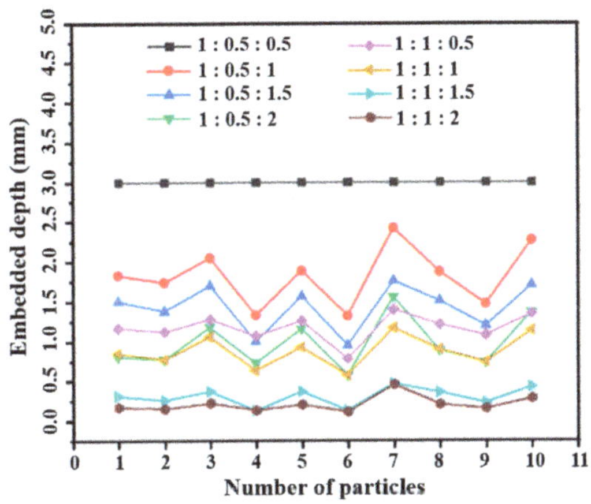

Figure 10. Embedding depths of aggregates in thin layers with different ratios.

Owing to the large discreteness of the measurement results, to reduce the influence and variability caused by test errors, the maximum and minimum values of the embedding depth of 10 aggregates under each group of cementitious material schemes were removed, and the average value of the embedding depth of the remaining eight aggregates was taken to obtain the embedding depth of the aggregates in the corresponding ratio of cementitious materials, as shown in Table 3. The relationship between the viscosity of the cementitious material and the embedding depth of the aggregates is shown in Figure 11.

Table 3. Depth of embedment of aggregates in thin layers of different viscosities.

Ratio (Glue/Powder/Sand)	Rotational Viscosity (Pa·s)	Insertion Depth (mm)
1:0.5:0.5	0.88	3.00 (fully embedded)
1:0.5:1	1.73	1.78
1:0.5:1.5	3.11	1.41
1:0.5:2	9.64	0.80
1:1:0.5	3.28	1.21
1:1:1	8.82	0.87
1:1:1.5	18.80	0.32
1:1:2	45.00	0.21

The test results show that the embedding depth of the aggregates decreased with increasing viscosity of the cementitious material. When the rotational viscosity was less than 0.88 Pa·s, the spread aggregates were completely embedded in the thin layer. When the rotational viscosity was 1.73 Pa·s, the embedding depth of the aggregates was 1.78 mm, and the embedding depth accounted for 59.3% of the particle size of the aggregates. When the rotational viscosity was 3.11 Pa·s, the embedding depth of the aggregates was 1.41 mm, and the embedding depth accounted for 47% of the particle size of the aggregates. When the rotational viscosity was 9.64 Pa·s, the embedding depth of the aggregates was 0.8 mm, and the embedding depth accounted for 26.7% of the particle size of the aggregates. When the rotational viscosity exceeded 9.64 Pa·s, the influence of the viscosity of the cementitious material on the embedding depth of the aggregates was not obvious. When the rotational viscosity exceeded 18.8 Pa·s, the embedding depth of the aggregates was 0.32 mm, and the embedding depth accounted for 10.7% of the particle size of the aggregates. As the viscosity increased, the embedding effect of the aggregates gradually weakened and the aggregates eventually floated on the thin layer surface.

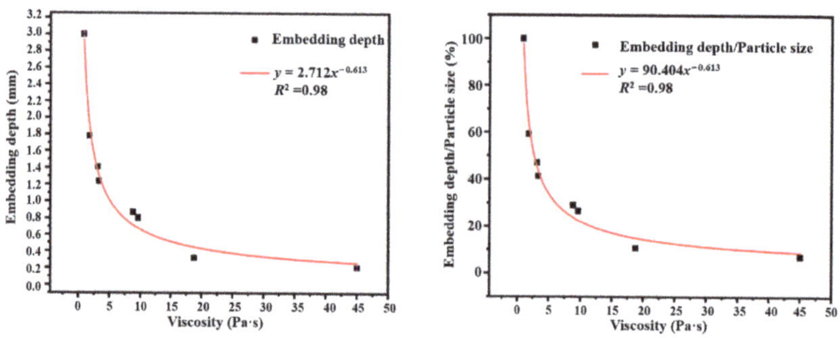

(**a**) Embedding depth (**b**) Percentage of embedded particle size

Figure 11. Embeddedness of aggregates in thin layers of different viscosities.

If the viscosity of the cementitious material is too small (less than 0.88 Pa·s), the aggregates are completely embedded in the thin layer, resulting in a lack of micro-convex bodies on the surface structure of the thin layer, and the spread aggregates cannot provide the anti-skid effect for the anti-skid thin layer. If the viscosity is too large (greater than 18.8 Pa·s), the embedding effect of the aggregates is not obvious, resulting in the aggregates only floating on the surface of the thin layer and poor anti-stripping performance.

Due to construction using the spreading method, in addition to the aggregates that first come into contact with the thin layer during the spreading being embedded in the thin layer to a certain depth, the excess aggregates covering the surface of the thin layer after spreading will press the lower layer of aggregates downward owing to their own weight. At this time, the embedding depth of the lower layer of aggregates will continue to increase until the cementitious material solidifies and stops sinking. Thus, when the viscosity of the cementitious material is 3.11 Pa·s, although the embedding depth of the aggregates accounts for 47% of the particle size of the aggregates and does not reach half the particle size of the aggregates, in actual working conditions, the lower layer of aggregates will continue to sink under the influence of the upper layer of aggregates, making its embedding depth exceed half the particle size of the aggregates. Thus, when the viscosity of the cementitious material is 1.73–3.11 Pa·s, the embedding depth of the aggregates is between 1/2 and 2/3 of the particle size.

In conclusion, the higher the viscosity of the cementitious material, the shallower the embedding of the aggregates. It is recommended that the viscosity of the cementitious material be controlled within 1.73–3.11 Pa·s. Within this range, the embedding depth of the aggregates is between 1/2 and 2/3 of the particle size, which can give full play to the role of the thin layer in dispersing stress, and the aggregates are not completely embedded in the thin layer, which can provide micro-convex bodies for the surface of the thin layer and improve the anti-skid performance of the thin layer.

3.3. Embedding Depth of Aggregates with Different Shape Characteristics

3.3.1. Shape Characteristics of Aggregates

The average values of the shape characteristic parameters of the aggregates are shown in Figure 12. It can be seen from the test results that the relationship of the axial coefficients of the three types of aggregates is in the order of C > B > A, among which the contour shape of type C aggregates is closer to needle-like. The relationship between the roundness factors of the three types of aggregates is A > B > C, among which the contour shape of type A aggregates is closer to circular. The relationship between the prism factors of the three types of aggregates is A > B > C, among which the contour change of type A aggregates is more regular.

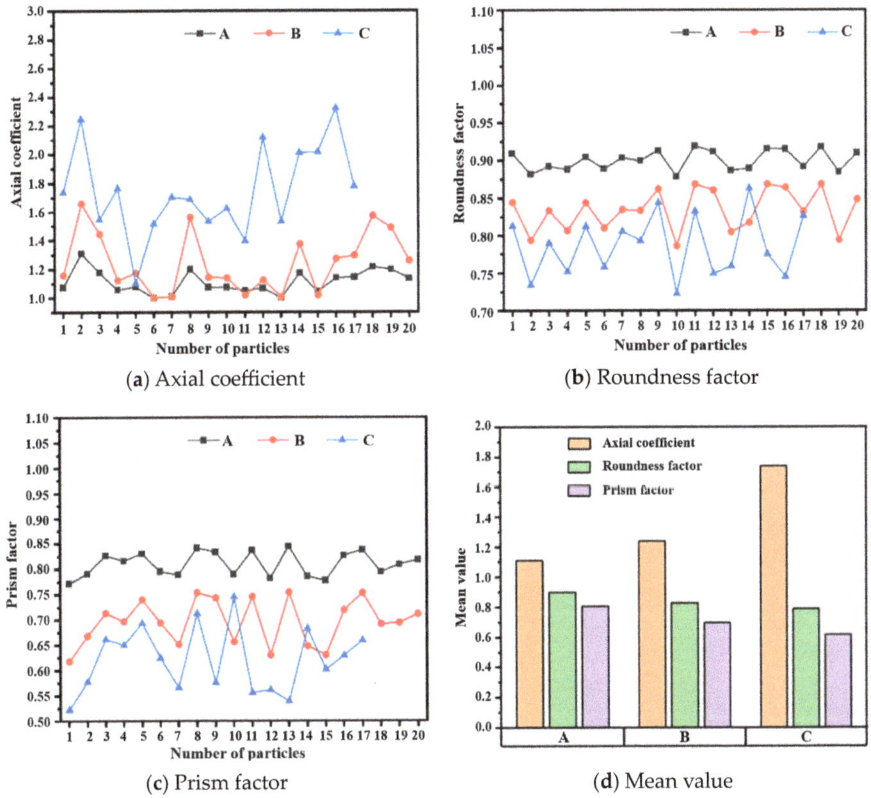

Figure 12. Characteristic parameters of aggregate shape.

3.3.2. Influence of Aggregate Shape on Embedding Depth

The three types of aggregates in Section 3.3.1 were spread into the cementitious material with an adhesive/powder/sand ratio of 1:0.5:1.5 and viscosity of 3.11 Pa·s according to the test method in Section 2.1, and their average embedding depth was measured. The test results are shown in Figure 13.

The test results showed that the shape characteristics of the aggregates have a significant impact on the embedding depth. The embedding depth was in the order of A > B > C. The average embedding depth of type A aggregates is 1.41 mm, that of type B aggregates is 1.29 mm, and that of type C aggregates is 0.62 mm. The difference in the embedding depth between types A and B is small, and the embedding depth of type C is the smallest.

The embedding depth of type A aggregates is the largest, with an average axial coefficient of 1.11, an average roundness factor of 0.9 (closest to 1), a shape closer to a circle, and average prism factor of 0.81, and a contour more regular and smoother. When the spread aggregates come into contact with the thin layer, because the shape of type A aggregates is rounder, the force can be concentrated in a smaller area, which is conducive to embedding the aggregates. During the embedding process, because the contour is more regular, the embedding friction resistance received will be smaller, and the embedding will be deeper. When type A aggregates are spread on the thin layer, nearly half of the part is embedded in the thin layer, and the thin layer can bond and wrap it to provide anti-stripping ability.

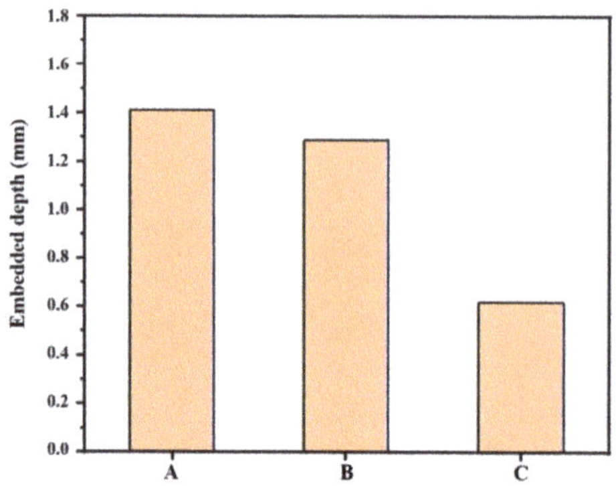

Figure 13. Depth of aggregate embedment.

The embedding depth of type B aggregates was slightly smaller than that of type A aggregates, with an average axial coefficient of 1.24, an average roundness factor of 0.83, and an average prism factor of 0.69. The axial coefficient was relatively small compared to that of type A aggregates. Its roundness was smaller than that of type A aggregates. When the spread aggregates come into contact with the thin layer, the area where the force is concentrated is larger than that of type A aggregates; therefore, the embedding effect is slightly worse than that of type A aggregates. Simultaneously, the prism factor of type B aggregates is smaller than that of type A aggregates. During the embedding process, owing to the irregular contour change, it receives greater embedding friction resistance, and the embedding depth will be smaller than that of type A aggregates.

The embedding depth of type C aggregates is the smallest, with an average axial coefficient of 1.74, an average roundness factor of 0.79, and an average prism factor of 0.62. This was a needle-shaped aggregate. When the aggregates are spread on the thin layer, they are mostly in a flat state to maintain stability. It has a large contact area with the thin layer, and its narrow and long structure makes it difficult for the resin mortar to climb, resulting in the smallest embedding depth of type C aggregates in the thin layer. The anti-stripping ability is mainly provided by the surface bonding with the thin layer material.

In conclusion, type B aggregates have a better embedding depth and prism. Considering both anti-skid and anti-stripping performance, type B aggregates should be preferred, and type C aggregates should be avoided.

3.3.3. Gray Relational Analysis of Aggregate Shape and Embedding Depth

Tables 4–6 present the initial data of the characteristic sequence $[X_1, X_2, X_3]$ and mother sequence X_0, dimensionless processing of the initial data, and correlation coefficient and correlation degree, respectively.

Table 4. Initial data of feature sequences and parent sequences.

Name	Factor	A	B	C
Insertion depth (mm)	X_0	1.41	1.29	0.62
Axial coefficient	X_1	1.11	1.24	1.74
Roundness factor	X_2	0.90	0.83	0.79
Prism factor	X_3	0.81	0.79	0.62

Table 5. Dimensionless processing of initial data of feature sequence and parent sequence.

Name	Factor	A	B	C
Insertion depth (mm)	X_0	1	0.915	0.440
Axial coefficient	X_1	1	1.117	1.568
Roundness factor	X_2	1	0.922	0.878
Prism factor	X_3	1	0.975	0.765

Table 6. Number of links and degree of association.

Shape Characterization Index	Factor	Correlation Coefficient			Relatedness
		A	B	C	
Axial coefficient	X_1	1	0.7361	0.3333	0.690
Roundness factor	X_2	1	0.9872	0.5628	0.850
Prism factor	X_3	1	0.8994	0.6339	0.844

Figure 14 shows the degree of correlation between the three factors and the embedding depth change rate. The correlation degree was obtained using the average value of the correlation coefficient, and the correlation degree ranged between 0 and 1. The larger the value, the stronger the correlation between the characteristic sequence and mother sequence.

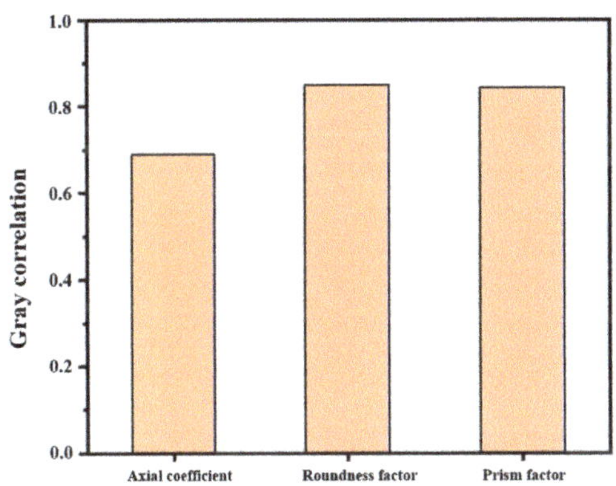

Figure 14. Gray correlation of shape feature indices.

Through gray relational analysis, it is concluded that the order of the correlation degrees of each shape characteristic index is roundness factor > prism factor > axial coefficient; that is, the roundness factor has the greatest correlation with the embedding depth and most obvious influence.

4. Conclusions

(1) An epoxy resin mortar layer with a composite filler of powder and sand provides better aggregate embedding than a single filler. This finding has direct implications for pavement construction, where proper aggregate embedding is critical for ensuring surface texture and bonding strength. The recommended filler ratios (adhesive-to-powder-to-sand at 1:0.5:1 or 1:0.5:1.5) are optimal for achieving a balance between viscosity and workability. If the ratio is exceeded, the viscosity rises sharply, reducing the construction performance. This means the mortar may become too stiff, leading to difficulty in spreading and reduced bonding with the aggregates. In practical terms, these factors can negatively impact pavement durability, skid resistance, and surface roughness, potentially increasing slipperiness and reducing long-term performance.

(2) The embedding depth of the aggregates is a crucial parameter for achieving the desired pavement characteristics. This study identifies that low viscosities below 0.88 Pa·s result in full embedding, which is beneficial for achieving smooth surfaces but may compromise the surface texture and skid resistance. High viscosities above 18.8 Pa·s cause aggregates to float on the surface, reducing structural bonding and durability. The identified optimal viscosity range of 1.73 to 3.11 Pa·s ensures a partial embedding of aggregates (1/2 to 2/3 of their particle size), providing a balance between an adequate surface texture for skid resistance and sufficient bonding for durability. This range is particularly useful in designing ultra-thin layers where spalling and wear resistance are critical.

(3) This study demonstrates that aggregates with varying shape characteristics influence their embedding behavior, which is essential for designing pavements with specific performance goals. For instance, Type A aggregates (7.07 mm^2) with high roundness and prism factors achieve the greatest embedding depth, making them suitable for applications requiring enhanced bonding and smoother surfaces. Type C aggregates (15.24 mm^2), with high axial coefficients but lower embedding depth, are more suitable for creating textured surfaces that improve skid resistance. The gray correlation analysis confirms that the roundness factor is the most critical parameter influencing embedding depth, suggesting that selecting aggregates with the appropriate shape characteristics can optimize skid resistance, roughness, and durability in pavement applications.

Author Contributions: Conceptualization, J.Y., G.Z., C.F. and C.L.; methodology, W.D. and Y.Z.; software, H.H.; validation, Y.Z., W.J. and L.X.; formal analysis, Y.Z.; investigation, H.H.; resources, G.Z.; data curation, J.Y. and H.L.; writing—original draft preparation, J.Y., H.L. and X.Y.; writing—review and editing, G.Z., W.D. and C.F.; supervision, G.Z., C.F. and C.L.; project administration, G.Z.; funding acquisition, G.Z. and C.L. All authors have read and agreed to the published version of the manuscript.

Funding: This research was funded by the Chongqing Natural Science Foundation Innovation and Development Joint Fund (CSTB2024NSCQ-LZX0076) and the Chongqing Jiaotong University Municipal Graduate United Cultivating Base (JDLHPYJD2023001).

Data Availability Statement: The original contributions presented in the study are included in the article, further inquiries can be directed to the corresponding authors.

Acknowledgments: The authors thank Chongqing Jiaoda Construction Engineering Quality Test Center Co., Ltd. For their help in sample testing.

Conflicts of Interest: Authors Jiaquan Yuan, Wenhong Duan, Weihong Jiang, Li Xiong, Huimei Li and Xiaohua Yang were employed by the company Dali Danan Highway Co., Ltd. The remaining authors declare that the research was conducted in the absence of any commercial or financial relationships that could be construed as a potential conflict of interest.

References

1. Shen, K.; Wang, H. Impact of dynamic loading on pavement deflection measurements from traffic speed deflectometer. *Measurement* **2023**, *217*, 113086. [CrossRef]
2. Yu, J.; Chen, F.; Deng, W.; Ma, Y.; Yu, H. Design and performance of high-toughness ultra-thin friction course in south China. *Constr. Build. Mater.* **2020**, *246*, 118508. [CrossRef]
3. Luo, H.; Chen, S.; Zheng, Y.; Wu, X.; Chen, C.; Huang, X. Improving skid-resistance durability of ultra-thin friction course in asphalt pavements through recycled steel slags as basalt replacements. *Constr. Build. Mater.* **2024**, *426*, 136138. [CrossRef]
4. Lou, K.; Xiao, P.; Wu, B.; Kang, A.; Wu, X.; Shen, Q. Effects of fiber length and content on the performance of ultra-thin wearing course modified by basalt fibers. *Constr. Build. Mater.* **2021**, *313*, 125439. [CrossRef]
5. Li, J.; Huan, X.; Wang, S.; Sheng, Y.; Xu, D.; You, Z. Performance of Optimized Composition of Epoxy Resin Adhesive Used in High Friction Surface Treatment. *Case Stud. Constr. Mater.* **2024**, *21*, e03431. [CrossRef]
6. Geng, L.; Ma, T.; Zhang, J.; Huang, X.; Hu, P. Research on performance of a dense graded ultra-thin wearing course mixture. *Appl. Sci.* **2017**, *7*, 800. [CrossRef]

7. Yu, J.; Feng, Z.; Chen, Y.; Yu, H.; Korolev, E.; Obukhova, S.; Zhou, J.; Zhang, Y. Investigation of cracking resistance of cold asphalt mixture designed for ultra-thin asphalt layer. *Constr. Build. Mater.* **2024**, *414*, 134941. [CrossRef]
8. Coni, M. Ultrathin multi-functional overlay. In *Airfield and Highway Pavement 2013: Sustainable and Efficient Pavements*; Al-Qadi, I.L., Murrell, S., Eds.; ASCE: Los Angeles, CA, USA, 2013; pp. 662–676.
9. Liu, Y.; Qian, Z.; Shi, X.; Zhang, Y.; Ren, H. Developing cold-mixed epoxy resin-based ultra-thin antiskid surface layer for steel bridge deck pavement. *Constr. Build. Mater.* **2021**, *291*, 123366. [CrossRef]
10. Li, X.; Ye, J.; Badjona, Y.; Chen, Y.; Luo, S.; Song, X.; Zhang, H.; Yao, H.; Yang, L.; You, L.; et al. Preparation and performance of colored Ultra-Thin overlay for preventive maintenance. *Constr. Build. Mater.* **2020**, *249*, 118619. [CrossRef]
11. McNeil, N.; Monsere, C.; Dill, J. Evaluation of driver comprehension and compliance of red-colored pavement markings for transit lanes in Portland, Oregon. *Transp. Res. Rec.* **2024**, *2678*, 123–137. [CrossRef]
12. Badin, G.; Ahmad, N.; Ali, H.M.; Ahmad, T.; Jameel, M.S. Effect of addition of pigments on thermal characteristics and the resulting performance enhancement of asphalt. *Constr. Build. Mater.* **2021**, *302*, 124212. [CrossRef]
13. Liu, H.; Zhang, Z.; Guo, D.; Peng, L.; Bao, Z.; Han, W. Research progress and prospect of application technology of thin-layer antiskid colored pavement at home and abroad. In Proceedings of the International Conference on Electric Technology and Civil Engineering (ICETCE), Lushan, China, 22–24 April 2011.
14. Liu, F.; Qu, D.; Tan, Z.; Yang, C.; Liu, Y. Experimental investigation of the effects of four anti-slide particles on the operational performance of coated coloured pavement. *Procedia Eng.* **2016**, *161*, 589–594. [CrossRef]
15. Lee, H.; Kim, Y. Laboratory evaluation of color polymer concrete pavement with synthetic resin binder for exclusive bus lanes. *Transp. Res. Rec.* **2007**, *1991*, 124–132. [CrossRef]
16. Wang, J.; Li, Q.; Song, G.; Luo, S.; Ge, D. Investigation on the comprehensive durability and interface properties of coloured ultra-thin pavement overlay. *Case Stud. Constr. Mater.* **2022**, *17*, e01341. [CrossRef]
17. Henderson, R.; Cook, G.; Cenek, P.; Patrick, J.; Potter, S. *The Effect of Crushing on the Skid Resistance of Chipseal Roads*; Land Transport New Zealand: Wellington, New Zealand, 2006.
18. Hu, L.; Yun, D.; Gao, J.; Tang, C. Monitoring and optimizing the surface roughness of high friction exposed aggregate cement concrete in exposure process. *Constr. Build. Mater.* **2020**, *230*, 117005. [CrossRef]
19. Yan, C.; Li, Q.; Wang, J.; Yang, H.; Wu, Y. Evaluation for long-term skid resistance of ultra-thin asphalt overlay based on texture characteristics. *Constr. Build. Mater.* **2024**, *438*, 137151. [CrossRef]
20. Jiang, W.; Yuan, D.; Shan, J.; Ye, W.; Lu, H.; Sha, A. Experimental study of the performance of porous ultra-thin asphalt overlay. *Int. J. Pavement Eng.* **2022**, *23*, 2049–2061. [CrossRef]
21. Ding, L.; Wang, X.; Zhang, K.; Zhang, M.; Yang, J.; Chen, Z. Durability evaluation of easy compaction and high-durability ultra-thin overlay. *Constr. Build. Mater.* **2021**, *302*, 124407. [CrossRef]
22. Chen, S.; Liu, X.; Luo, H.; Yu, J.; Chen, F.; Zhang, Y.; Ma, T.; Huang, X. A state-of-the-art review of asphalt pavement surface texture and its measurement techniques. *J. Road Eng.* **2022**, *2*, 156–180. [CrossRef]
23. Roy, N.; Mondal, P.G.; Kuna, K.K. Image-based indices of aggregates for predicting the initial skid resistance of bituminous pavements. *Constr. Build. Mater.* **2023**, *400*, 132776. [CrossRef]
24. Lin, C.; Wang, T. Effect of fine aggregate angularity on skid-resistance of asphalt pavement using accelerated pavement testing. *Constr. Build. Mater.* **2018**, *168*, 41–46. [CrossRef]
25. Kim, Y.R.; Aragao, F.T.; Allen, D.H.; Little, D.N. Damage modeling of bituminous mixtures considering mixture microstructure, viscoelasticity, and cohesive zone fracture. *Can. J. Civ. Eng.* **2010**, *37*, 1125–1136. [CrossRef]
26. Boulangé, L.; Bonin, E.; Saubot, M. Physicochemical characterisations of the bitumen–aggregate interface to get a better understanding of stripping phenomena. *Road Mater. Pavement Des.* **2013**, *14*, 384–403. [CrossRef]
27. Wang, F.; Li, Y., Yu, L.; Pang, W. Study on influencing factors of asphalt-aggregate stripping mechanism. *Adv. Mater. Sci. Eng.* **2021**, *2021*, 6619118. [CrossRef]
28. Boz, I.; Kumbargeri, Y.S.; Kutay, M.E. Performance-based percent embedment limits for chip seals. *Transp. Res. Rec.* **2019**, *2673*, 182–192. [CrossRef]
29. Kumbargeri, Y.; Boz, I.; Kutay, M.E.; Heidelberg, A. A study on the effects of aggregate shape and percent embedment on chip seal performance via image-based finite element analysis. *Int. J. Pavement Eng.* **2020**, *21*, 1002–1011. [CrossRef]
30. Ozdemir, U.; Kutay, M.E.; Hibner, D.; Lanotte, M.; Kumbargeri, Y.S. Quantification of aggregate embedment in chip seals using image processing. *J. Transp. Eng. Part B Pavements* **2018**, *144*, 04018047. [CrossRef]
31. Seitllari, A.; Kutay, M.E. Soft computing tools to predict progression of percent embedment of aggregates in chip seals. *Transp. Res. Rec.* **2018**, *2672*, 32–39. [CrossRef]
32. Ma, Z.; Wang, H.; Zhang, Y.; Jelagin, D.; Li, Y. Effects of aggregate shape on the macrotexture and performance of chip seal: A laboratory study. *Road Mater. Pavement Des.* **2024**, 1–22. [CrossRef]
33. *GB/T 22314-2008*; Plastics-Epoxy Resins-Determination of Viscosity. Standards Press of China: Beijing, China, 2008.

buildings

MDPI

Article

Incorporating Recycled Textile Fibers into Stone Mastic Asphalt

Carlos Alonso-Troyano *, David Llopis-Castelló and Blanca Olaso-Cerveró

Highway Engineering Research Group, Universitat Politècnica de València, Camino de Vera s/n, 46022 Valencia, Spain; dallocas@upv.es (D.L.-C.); bolacer@upv.edu.es (B.O.-C.)
* Correspondence: caraltro@upv.es

Abstract: The increasing environmental impact of industrial waste, particularly from the textile sector, has driven efforts to integrate alternative materials into road construction. This study explores the feasibility of incorporating recycled cotton textile fibers into Stone Mastic Asphalt (SMA) mixtures to enhance their mechanical performance and sustainability. The bituminous mixture SMA 11 surf 35/50 was designed with 0.3% textile fibers, a dosage optimized to prevent binder drainage while maintaining adequate structural properties. Laboratory tests were conducted to evaluate bulk and maximum density, air void content, water sensitivity, and resistance to permanent deformation. The results demonstrated that the inclusion of 0.3% textile fibers significantly reduced binder drainage, improved moisture resistance with an ITSR of 96.30%, and enhanced stability under traffic loads. Although the WTSAIR value of 0.12 mm/1000 cycles did not fully comply with PG-3 requirements for T2 traffic, slight adjustments in binder content or composition could optimize performance. Beyond technical benefits, this study highlights the environmental and economic advantages of repurposing locally generated textile waste, reducing landfill accumulation, and fostering synergies between industries. Future research should focus on optimizing bitumen content, conducting fatigue and aging tests, and validating field performance under real traffic and environmental conditions to ensure long-term durability and compliance with road specifications.

Keywords: asphalt mixture; recycled textile fibers; sustainable material; pavement; stone mastic asphalt

1. Introduction

Transport, particularly the road sector, has a significant environmental impact due to its energy consumption and greenhouse gas emissions. Authorities must address this challenge through sustainable practices in road construction and maintenance. The European Union promotes the circular economy, which includes the reuse and recycling of waste materials, such as construction and demolition waste (CDW) and textile waste.

To enhance environmental quality and promote the sustainable development of transport infrastructure, it is essential to implement sustainable strategies in pavement construction and rehabilitation. The benefits of using large quantities of recycled materials include (i) the reduction in natural resource consumption, (ii) the elimination of waste materials otherwise destined for disposal, (iii) the reduction in energy and water consumption, and (iv) the decrease in greenhouse gas emissions [1].

However, the textile industry is one of the most polluting sectors, generating a significant amount of waste. This waste is produced not only during garment manufacturing but also during the dyeing process. To prevent textile industry waste from ending up in landfills, one of the key innovations of Directive 2018/851/EU is that companies will be required to pay for the waste they generate and will no longer be allowed to destroy or incinerate it.

In the field of road construction, research has explored the potential use of various types of waste as substitutes for natural aggregates in pavement and granular layers and even as a partial replacement for bitumen. In this context, one of the most commonly reused materials is reclaimed asphalt pavement (RAP) [2], along with construction and demolition waste (CDW) [3,4]. Other waste materials that have been studied include those from the steel industry [5], plastics [6], and rubber [7], which have shown potential to improve performance or reduce environmental impact. Other waste materials that have been studied include those from the steel industry and plastics. However, very few studies have assessed the benefits of incorporating textile waste into bituminous mixtures, which could enhance durability by contributing to a more resilient structural framework.

Since the late 19th century, asphalt has been widely used in road construction [8]. SMA mixtures, introduced in Germany in the 1960s, are recognized for their high performance and durability, making them ideal for heavy traffic pavements [9]. These mixtures offer fatigue resistance and long-term stability [9], requiring a higher asphalt binder content (6–7%) compared to conventional mixtures (4–7%) [10].

Although SMA mixtures have a higher initial cost, their long-term benefits—such as enhanced durability, resistance to studded tires, wear, and plastic deformation—justify their widespread adoption worldwide, especially for high-traffic roads. Despite certain challenges like binder exudation, additional advantages including improved drainage and noise reduction make SMA mixtures a cost-effective solution in the long run [9,11–13].

Fibers play a fundamental role in Stone Mastic Asphalt (SMA) mixtures, as they stabilize the binder and prevent segregation. In addition to enhancing mechanical performance by increasing resistance to fatigue and deformation, they also contribute significantly to reducing cracking and help delay and control the appearance of reflective cracks [14]. This reinforces the suitability of SMA as an effective solution for high-traffic roads. Different additives such as cellulose, polyester, lignin, and glass fibers have been studied for their ability to enhance the dynamic modulus, moisture tolerance, and fatigue resistance of SMA, while also reducing binder segregation [15–17]. Studies have shown that cellulose fiber, when added in proportions of approximately 0.30% by weight of the mixture, significantly reduces drainage effects and improves mechanical properties [18].

Numerous studies have demonstrated that the incorporation of cellulose fibers into SMA mixtures significantly enhances their mechanical properties and durability. These fibers, typically added in proportions around 0.3% by weight of the total mixture, are effective in stabilizing the binder and preventing drainage due to their high absorption capacity and structural integrity. For instance, Putman and Amirkhanian [19] reported over 30% reduction in binder drainage and significant improvements in resistance to permanent deformation upon using cellulose fibers. Similarly, Behbahani et al. [20] observed reductions of approximately 30% in rutting compared to mixtures without fibers. Additional research by Qian and Lu [21] confirmed improvements in indirect tensile strength and binder cohesion. These findings underscore the functional role of fibers in improving performance under traffic loads and environmental stresses. Despite these advances, most existing research focuses on commercial cellulose fibers. There is a clear gap in assessing alternative fibrous materials—such as recycled textile waste—that may offer comparable stabilizing effects while contributing to environmental sustainability, thereby justifying the present investigation.

Hybrid modifications of SMA using different fibers have yielded promising results. For instance, cellulose fibers improve exudation capacity, ductility, and fatigue resistance, while basalt fibers reduce permanent deformation and increase stress resistance [22]. Other studies have tested alternative fibers such as palm oil-derived cellulose, which, when combined with ground tire rubber, enhances pavement performance [23]. Recycled tire textile fibers (WTTF) have also been investigated as a replacement for commercial additives, demonstrating similar mechanical properties to traditional SMA blends [15,24].

Although various materials have been studied to absorb excess binder in SMA mixtures, including cellulose, basalt, pumice, paper, palm oil fibers, and synthetic fibers such as tire textiles and cigarette filters, there is limited knowledge regarding the use of textile waste fibers in SMA mixtures [25].

The textile industry generates a wide variety of waste throughout its production chain, including fiber remnants, fabric scraps, threads, and byproducts from textile recycling. Among these, cotton fibrous waste generated during the mechanical cutting and processing of recycled textiles stands out. This type of waste, which is homogeneous in nature and composed exclusively of natural fibers, has a length ranging from 0 to 5 mm, making it particularly suitable for use as an additive in bituminous mixtures.

In this study, a cotton fibrous waste discarded by the textile industry itself—due to its non-reusability in spinning processes—was selected. Its choice is based both on technical suitability—given its ability to stabilize the bituminous binder and reduce drainage—and on sustainability criteria, as it is a locally available waste that currently lacks a valorization pathway. Its incorporation into SMA-type mixtures contributes to a more sustainable paving model, promoting the reuse of industrial waste and reducing reliance on virgin raw materials.

Previous research suggests that local materials could replace cellulose fibers, but further investigation is required. In this context, the Valencian Community, a key textile production region in Spain, generates approximately 150,000 tonnes of textile waste annually, with cities like Alcoy, Ontinyent, and Elche being major contributors. Alcoy alone reports that 25% of its industrial waste is textile-based. In response, the region has launched initiatives such as the ReFashion project, promoted by the Instituto Valenciano de Competitividad Empresarial (IVACE), to recycle and reuse textile materials, reducing landfill disposal. Additionally, local companies are adopting more sustainable technologies, creating approximately 500 new jobs in the last five years.

One promising solution is the incorporation of textile waste into asphalt mixtures, which can enhance mechanical properties, increase durability, and reduce maintenance costs. By integrating textile waste into asphalt, the demand for virgin materials decreases, lowering environmental impact while promoting a circular economy. Investing in innovative recycling strategies and fostering collaboration between government, businesses, and society are crucial steps toward a more sustainable textile industry.

Objective and Contribution

The primary objective of this study is to analyze the feasibility of using textile waste fibers as an alternative to commercial cellulose fibers in SMA mixtures. This includes defining optimal component proportions and conducting laboratory tests to evaluate the suitability of the formulated mixtures. The bituminous mixtures were characterized through volumetric and performance-based tests, assessing fundamental properties such as void content, water sensitivity, and resistance to permanent deformation.

By incorporating textile waste fibers into SMA, this study not only explores a novel recycling approach but also contributes to a circular economy by reducing waste disposal and improving pavement durability. Additionally, this research aligns with sustainability goals by reducing reliance on virgin materials and promoting the reuse of industrial byproducts.

Therefore, this paper investigates the potential of using textile waste fibers as a sustainable additive in Stone Mastic Asphalt (SMA) mixtures, aiming to evaluate their effectiveness as a substitute for commercial cellulose fibers. The study includes a comprehensive experimental program to assess the mechanical performance of the modified mixtures. The findings contribute to advancing circular economy strategies in pavement engineering while addressing a significant environmental challenge posed by the textile industry.

2. Materials and Methods

2.1. Materials

The selected asphalt mixture for this study is SMA 11 surf 35/50, which was modified with recycled cotton fibers to obtain the final product. The composition of the aggregates used in the mixture are as follows: (i) filler or mineral powder smaller than 0.063 mm; (ii) a 0/6 mm fraction consisting of fine limestone aggregate; and (iii) a 6/12 mm fraction of coarse porphyritic aggregate. Upon receipt of the aggregates in the laboratory, granulometric tests were conducted following the UNE-EN 933-2 [26] standard to determine the particle size distribution (see Table 1). The aggregates were supplied by Guerola Áridos y Hormigones, S.L., based in Ontinyent, Alicante (Spain).

Table 1. Particle size distribution of materials.

Type of Aggregate	Test Sieves for Aggregates UNE-EN 933-2 (mm)							
	22	16	11.2	8	4	2	0.5	0.063
Filler	100	100	100	100	100	100	98.2	75.3
0/6 Fraction	100	100	100	98.93	76.67	66.77	38.74	3.15
6/12 Fraction	100	100	89.58	34.76	0.45	0.24	0.22	0.18

The bitumen used in the production of the SMA 11 mixture was 35/50 penetration grade, supplied by the Spanish company Cepsa. This bitumen meets the required properties for a wearing course up to a traffic category T2 (AADT < 800 veh/day), according to Spanish road specifications (PG-3) [10]. The main characteristics of the bitumen are presented in Table 2.

Table 2. Specifications of conventional bitumen 35/50.

Characteristics	Unit	Standard	Min	Max
Original Bitumen				
Penetration (25 °C, 100 g, 5 s)	0.1 mm	UNE-EN 1426 [27]	35	50
Softening point	°C	UNE-EN 1427 [28]	50	58
Penetration index	-	UNE-EN 12591 [29]	−1.5	0.7
Fraass breaking point	°C	UNE-EN 12593 [30]	−5	-
Solubility	%	UNE-EN 12592 [31]	99	-
Flash point	°C	UNE-EN 2592 [32]	240	-
Residue after thin-film and rotating film test				
Mass variation	%	UNE-EN 12607-1 [33]	-	0.5
Penetration (25 °C, 100 g, 5 s)	% of original	UNE-EN 1426 [27]	53	-
Softening point variation	°C	UNE-EN 1427 [28]	-	11

According to the manufacturer's recommendations, the working temperatures for the asphalt mixture are as follows (Figure 1):

- Mixing temperature: 155–165 °C;
- Binder working temperature: 155–165 °C;
- Compaction temperature: 150–160 °C;
- Maximum heating temperature: 170 °C.

Figure 1. Laboratory heated mixer.

The addition of recycled cotton textile fibers modifies the rheological properties of bitumen by increasing viscosity and enhancing resistance to binder drainage, thus improving overall mixture cohesion. Traditionally, fibers are incorporated into asphalt mixtures to minimize binder drainage. The fibers were provided by Recover Textile Systems, S.L., a company specializing in processing textile waste located in Banyeres de Mariola, Alicante (Spain). According to analyses conducted by this company, the cotton powder consists of 100% cotton fibers, ensuring the absence of harmful substances or compounds that could pose health risks. The fibrous particles in this waste ranged in length from 0 to 5 mm and originated exclusively from cutting residues generated during the manufacturing of new recycled cotton fibers (see Figure 2). In this study, the fibers used were approximately 5 mm in length. This controlled production process guarantees the homogeneity of the subproduct, making it a consistent and reliable material for asphalt applications. Recycled cotton fibers are specifically utilized due to their local availability, as the Valencian Community is a significant center for textile waste generation, particularly from cities such as Alcoy, Ontinyent, and Elche. This provides considerable logistical and environmental advantages by reducing transportation-related costs and minimizing the carbon footprint associated with material transportation. By employing these local wastes in asphalt mixtures, abundant regional resources are utilized, contributing to local economic development, enhancing synergies among nearby industries, and effectively promoting the implementation of a circular economy in the road construction sector.

Based on these materials and the granulometric spindles specified in Article 544 of the Spanish road specifications (PG-3) [10], which defines the composition of SMA 11 surf 35/50, the minimum bitumen content for the mixture is set at 5.8%. However, if the aggregate density differs from 2.65 g/cm^3, a correction factor ($\alpha = 2.65/\rho_d$) could be applied. Given that the aggregates supplied have a density of 2.9 g/cm^3, the adjusted minimum bitumen content is 5.3%.

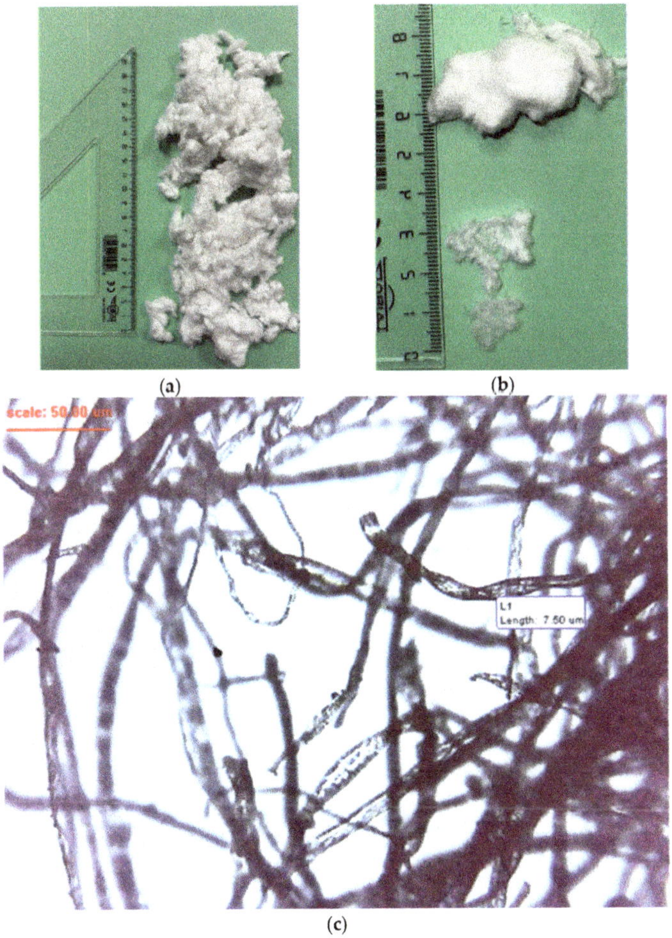

Figure 2. Recycled cotton textile fibers: (**a**) general photo, (**b**) detail photo, and (**c**) micrograph image.

Three bituminous mixtures were produced, incorporating 0.0%, 0.3%, and 0.6% cotton textile fibers. The proportions of aggregate and filler used in each mixture are detailed in Table 3 and Figure 3.

Table 3. Percentages of materials in the dosed mixtures.

Type of Asphalt Mixture	Filler < 0.063	Aggregate 6–12	Aggregate 6–0	Tex Fibers
SMA 11 0.0%	8%	65%	27%	0%
SMA 11 0.3%	8%	64.70%	27%	0.30%
SMA 11 0.6%	8%	64.40%	27%	0.60%

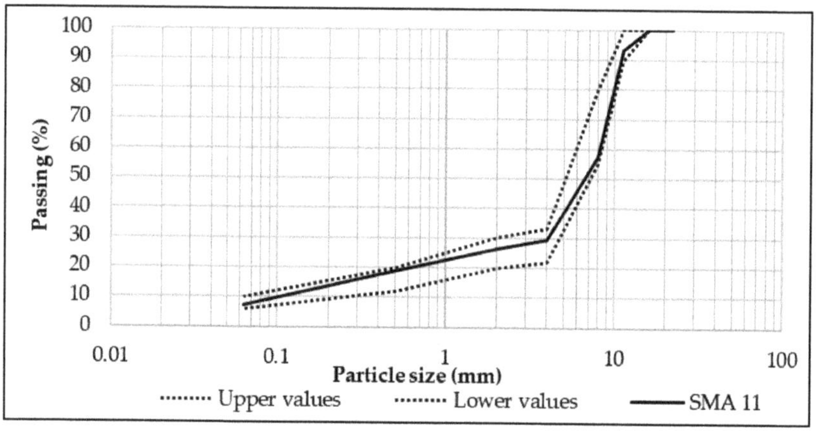

Figure 3. Particle size distribution.

2.2. Fabrication and Preparation of Test Specimens

First, the binder drainage test was carried out. Three samples were prepared, containing 0.0%, 0.3%, and 0.6% of cotton textile fibers, respectively.

For the selection of these fiber percentages, it has been considered that Article 544.2.4 of PG-3 [10] specifies that, in the case of mixtures using fibers, their content must not be lower than 0.3%. Therefore, we decided to test a mixture without fibers, a mixture with the minimum amount of fibers allowed by Spanish regulations, and a mixture with 0.6% fibers.

The preparation of eight cylindrical specimens (Ø 101.6 mm and height 63.5 mm) was carried out using impact compaction, applying 50 blows to each face, to determine the bulk density. Additionally, an extra amount of material was prepared to determine the maximum density. Based on the density values and the bitumen and additive content, the air void content of the mixture was estimated.

Subsequently, the water sensitivity test was conducted to determine the ITSR index. For this purpose, other eight cylindrical specimens were prepared and then divided into two batches: dry and wet.

Additionally, based on the obtained air void values, four prismatic specimens (410 mm × 260 mm × 40 mm) were fabricated using a plate compactor for the wheel tracking test, allowing for the determination of plastic deformations.

2.3. Characterization of the SMA

2.3.1. Binder Drainage

To determine the optimal textile fiber content, as specified in Article 544 of PG-3 [10], the binder drainage test was conducted in accordance with the UNE-EN 12697-18 [34] standard, using the beaker method.

The method consists of evaluating the amount of binder lost due to drainage after one hour at a representative temperature. To do this, the empty beaker (m_0) is weighed and, subsequently, a sample of the bituminous mixture is placed in the glass, also obtaining the overall weight (m_2). After one hour in the oven at the representative temperature, the glass is emptied by tipping the bituminous mixture material and the empty vessel (m_1) is weighed again. The difference between the weight of the beaker after (m_1) and before (m_0) corresponds to the amount of binder drained. Finally, the binder flow (BD) is expressed as a percentage using Equation (1).

$$BD = \frac{m_1 - m_0}{m_2 - m_0} \quad (1)$$

The test was carried out three times for each mixture, corresponding to the different textile fibers contents under evaluation.

2.3.2. Bulk Density, Maximum Density, and Air Voids

The bulk density of the SMA mixtures was determined following the UNE-EN 12697-6 [35] standard, using Method A—Immersion in water. This method is based on Archimedes' principle, where the specimen is submerged in water to determine its volume by displacement. The process consists of the following steps:

1. Sample preparation:
    - Drying: The specimen is dried at room temperature until a constant weight is achieved.
    - Weighing in Air: The dry specimen is weighed, recording its mass as m_1.
2. Heavy underwater:
    - The specimen is immersed in deaerated water at 25 ± 1 °C.
    - The submerged saturated mass is recorded as m_2.
    - The saturated surface-dry mass is recorded as m_3.
3. Calculation of bulk density:

$$\rho = \frac{m_1}{m_3 \quad m_2} \times \rho_W \tag{2}$$

where ρ is the bulk density (g/cm^3), m_1 is the mass of the dry specimen (g), m_2 is the mass of the saturated specimen immersed in water (g), m_3 is the mass of the saturated specimen (g), and ρ_W is the density of water at the test temperature (g/cm^3).

The maximum density of the asphalt mixtures was determined in accordance with the UNE-EN 12697-5 [36] standard. This parameter is essential for evaluating the compaction quality of asphalt mixtures and ensuring proper pavement performance, as it allows for the subsequent calculation of the air void content in the mixture.

The method used was Procedure A—Volumetric Procedure, which involves measuring the sample volume using a pycnometer filled with water. This approach provides an accurate determination of the maximum density by applying Archimedes' principle to measure the displaced water volume.

The maximum density was determined following a structured procedure to ensure accuracy. The process consists of the following steps:

1. Sample Preparation:
    - A representative sample of the bituminous mixture is selected.
    - The sample is dried at room temperature to eliminate any surface moisture.
2. Weighing the Sample:
    - The empty and dry pycnometer is weighed (m_1).
    - The test sample is placed inside the pycnometer, and the total mass is recorded (m_2).
    - The pycnometer is then filled with de-aired water up to the calibration mark, and the combined mass is recorded (m_3).
3. Calculation of Maximum Density:

$$\rho_{mv} = \frac{m_2 - m_1}{10^6 \times V_p - (m_3 - m_2)/\rho_W} \tag{3}$$

where ρ_{mv} is the maximum density (g/cm^3), m^1 is the mass of the empty pycnometer, m^2 is the mass of the pycnometer and the test sample (g), m_3 is the mass of the pycnometer, the test sample, and the water, V_p is the volume of the pycnometer (m^3), and ρ_W is the density of water at the test temperature (g/cm^3).

The air void content (V_a) in the mixture was determined according to the UNE-EN 12697-8 [37] standard. This parameter is calculated using the maximum density (ρ_{mv}) and the bulk density (ρ) of the specimens by applying Equation (4).

$$V_a = \frac{\rho_{mv} - \rho_b}{\rho_{mv}} \times 100 \quad (4)$$

Additionally, the void content in aggregates (*VMA*) and the void content filled with a binder (*VFB*) can be determined using the following expressions:

$$VMA = V_a + \frac{B \times \rho_b}{\rho_b} \quad (5)$$

$$VFB = \frac{B \times \rho \times 100}{\rho_b \times VMA} \quad (6)$$

where *Va* is the air void content in the bituminous specimen (%), *VFB* is the percentage of voids filled with binder (%), *VMA* is the void content in aggregates (%), ρ_{mv} is the maximum density of the mixture (g/cm^3), ρ is the bulk density of the specimen (g/cm^3), *B* is the binder percentage in the specimen (%), and ρ_b is the binder density (g/cm^3).

The results must comply with the limits established in Table 544.10 of PG-3 [10] to ensure the required compaction and durability of the SMA-type mixtures (see Table 4).

Table 4. Air void content limits for SMA-type bituminous mixtures according to PG-3 [10].

MIX TYPE	% AIR VOIDS (Standard UNE-EN 12697-8)
SMA 8	4–6
SMA 11	
SMA 16	4–7

2.3.3. Water Sensitivity Test

The water sensitivity of the bituminous mixtures was evaluated according to the UNE-EN 12697-12 [38] standard, using Method A, which is based on the UNE-EN 12697-23 [39] standard for determining the indirect tensile strength of bituminous specimens.

For this test, eight cylindrical specimens (Ø 101.6 mm, height 63.5 mm) were compacted using the impact method (UNE-EN 12697-30) [40], applying 50 blows per face. After fabrication, the specimens were divided into two groups: a dry group and a wet group. The difference in average apparent densities between both groups was required to remain below 0.015 g/cm^3 to ensure consistency in the results.

The dry specimens were stored upright on a flat surface at (20 ± 5) °C. The wet specimens underwent a vacuum saturation process and were subsequently immersed in a water bath at (40 ± 2) °C for (72 ± 2) hours.

The test temperature was set at 15 °C. To achieve this temperature, the dry specimens were placed in a thermally controlled air chamber, while the wet specimens were submerged in a water bath for at least four hours before testing.

Finally, the indirect tensile strength (*ITS*) of both specimen batches was measured according to the UNE-EN 12697-23 [39]. The indirect tensile strength ratio (*ITSR*), which quantifies the effect of moisture on the mechanical properties of the mixture, was calculated using Equation (7).

$$ITSR = \frac{ITS_w}{ITS_d} \times 100 \quad (7)$$

where $ITSR$ is the indirect tensile strength ratio, in percentage (%); ITS_w is the average indirect tensile strength of the wet batch, in kPa; and ITS_d is the average indirect tensile strength of the dry batch, in kPa.

2.3.4. Wheel Tracking Test

To evaluate the susceptibility of bituminous mixtures to deformation under load, the test defined in the UNE-EN 12697-22 [41] standard was conducted. This required the fabrication of four prismatic specimens with dimensions of 410 mm × 260 mm × 40–60 mm, following the UNE-EN 12697-33 [42] standard for specimen preparation using a plate compactor.

After manufacturing and compacting the mixture, it was allowed to cool before being demolded. The specimens were then stored on a flat surface for at least two days to ensure proper stabilization before testing.

The test was carried out using a testing machine where a loaded wheel repeatedly moves over a securely fixed pavement sample. It was performed following Procedure B in air, at a controlled temperature of 60 °C, with a total duration of 10,000 cycles.

3. Results

3.1. Binder Drainage

Figures 4 and 5 present the results obtained from the binder drainage test, conducted according to the UNE-EN 12697-18 [34] standard. As can be observed, the incorporation of recycled textile fibers significantly reduces binder drainage. With the addition of 0.3% fibers, drainage is almost entirely eliminated, as illustrated in Figure 4.

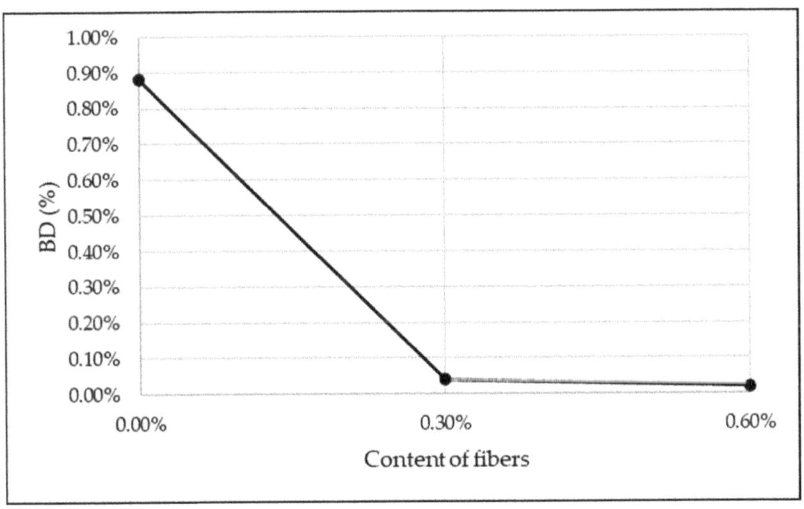

Figure 4. Binder drainage depending on content of fibers.

The results confirm that the addition of 0.3% recycled textile fibers to the bituminous mixture is effective in mitigating binder drainage and complies with the limits established in PG-3 [10]. Article 544.5.1.5 states that for SMA 11 mixtures, binder drainage, determined using the beaker method UNE-EN 12697-18 [34], must be below 0.3%, with a recorded value of 0.04%.

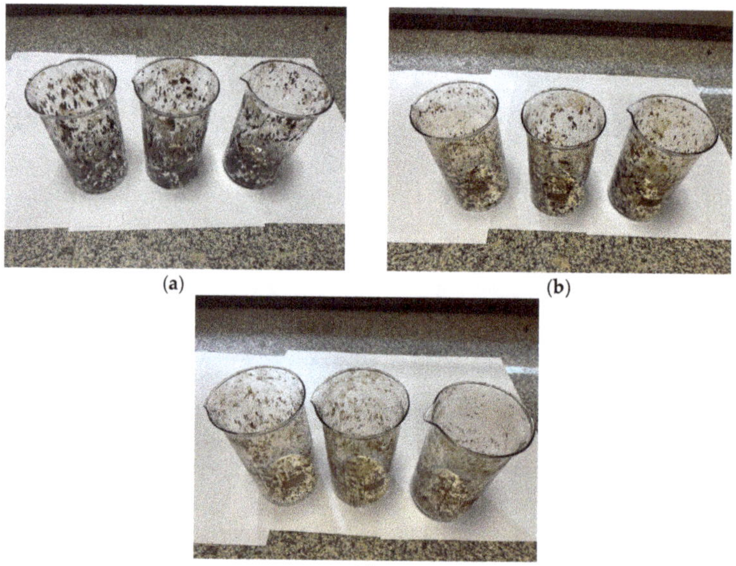

(a) (b) (c)

Figure 5. Results of the binder drainage test: (**a**) 0.0% fibers, (**b**) 0.3% fibers, and (**c**) 0.6% fibers.

3.2. Air Voids

The bulk density of the SMA 11 surf 35/50 mixture with 0.3% fibers was determined from eight specimens, obtaining an average value of 2.462 g/cm^3. The maximum density was calculated from two samples, yielding an average value of 2.602 g/cm^3.

Using these density values, the air void content (Va) was computed (see Table 5). On average, the SMA 11 surf 35/50 mixture with 0.3% fibers and a bitumen content of 5.3% exhibited an air void content of 5.4%, which falls within the 4% to 6% range specified in Article 544 of the PG-3 [10] for this type of mixture.

Table 5. Air void content of SMA 11 surf 35/50 with 0.3% fibers.

Specimen	ρ (g/cm^3)	Va (%)	VMA (%)	VFB (%)
I	2.469	5.124	17.767	71.161
II	2.467	5.193	17.827	70.869
III	2.464	5.315	17.932	70.363
IV	2.462	5.398	18.004	70.021
V	2.454	5.696	18.263	68.811
VI	2.467	5.197	17.830	70.855
VII	2.456	5.627	18.204	69.086
VIII	2.456	5.631	18.207	69.070
Average	2.462	5.398	18.004	70.029

Additionally, the voids in the mineral aggregate (VMA) and the voids filled with bitumen (VFB) were analyzed. The obtained values indicate that the mixture maintains the necessary balance between void structure and binder content, ensuring adequate performance and durability. The average VMA was 18.0%, while the VFB was 70.0%, confirming that the mix design meets the required compaction and stability criteria.

These results demonstrate that incorporating 0.3% recycled textile fibers contributes to an optimized air void content while maintaining compliance with road construction specifications.

3.3. Water Sensitivity Test

The results obtained from the water sensitivity test are presented in Figure 6, which illustrates the indirect tensile strength (ITS) values for both wet and dry specimens.

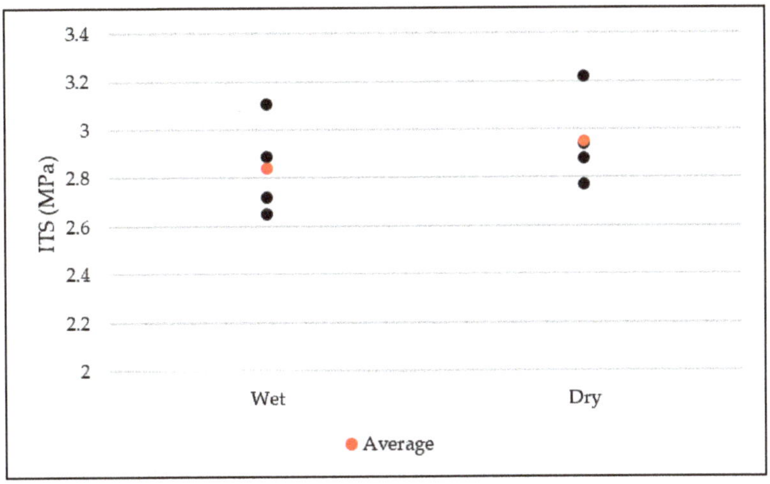

Figure 6. Results of the water sensitivity test.

The average ITS value for the dry specimens was 2.95 kPa, while the wet specimens achieved an average of 2.84 kPa. Using these values and applying the corresponding formula, the indirect tensile strength ratio (ITSR) was calculated to be 96.30%.

According to PG-3 [10], the ITSR value for this type of mixture must be greater than 90% to ensure sufficient resistance to moisture damage. The obtained ITSR of 96.30% confirms that the SMA 11 surf 35/50 mixture with 0.3% recycled textile fibers meets the required specifications, demonstrating excellent water resistance and mechanical stability under moisture exposure.

These results indicate that the incorporation of recycled textile fibers does not compromise the mixture's ability to withstand water-induced damage, making it a viable and durable alternative for road construction.

3.4. Wheel Tracking Test

The results obtained from the wheel tracking test are presented in Figure 7 and Table 6. According to Article 544 of the PG-3 [10], an SMA-type mixture intended for surface courses must have a WTSAIR value lower than 0.10 mm/1000 cycles for moderate and low heavy traffic categories ($\leq$T2) and a lower than 0.07 mm/1000 cycles for significant heavy traffic categories ($\geq$T1).

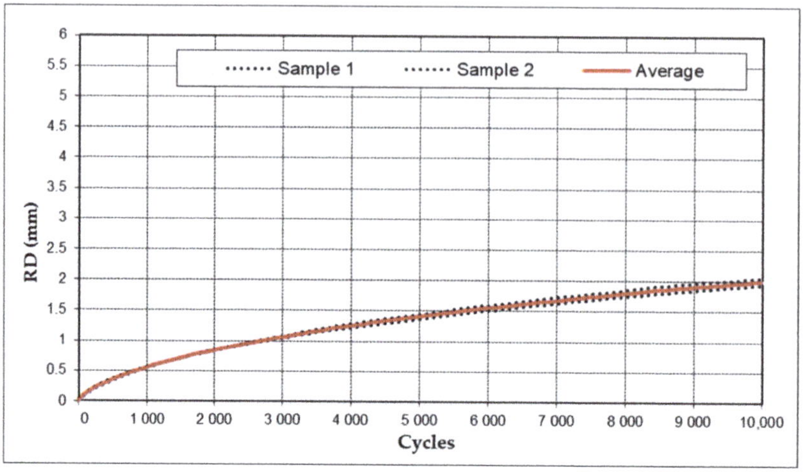

(a)

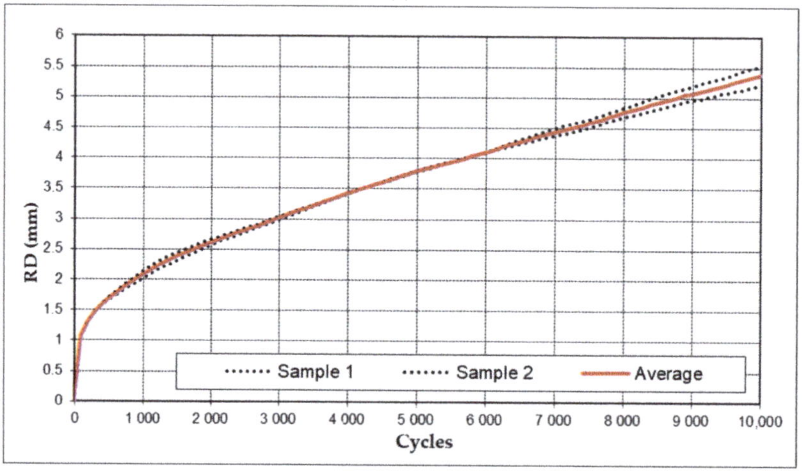

(b)

Figure 7. Track test results for SMA: (**a**) without fibers and (**b**) with 0.3% fibers.

Table 6. Track test results.

Sample	WTS_{AIR}
SMA 11 without fibers	0.32
SMA 11 with 0.3% fibers	0.12

The SMA 11 mixture without fibers exhibited a WTSAIR value of 0.32 mm/1000 cycles, while the mixture containing 0.3% recycled textile fibers achieved a significantly lower value of 0.12 mm/1000 cycles. Although this value does not fully comply with the PG-3 requirements [10], it is close to the specified threshold.

These results suggest that further optimization of the mix design, such as adjusting the binder content or modifying the binder type, could enhance the resistance to permanent deformation, ensuring full compliance with the standard. Despite not meeting the exact limit, the significant reduction in rutting potential with fiber incorporation highlights the positive effect of recycled textile fibers in improving the mixture's performance.

4. Discussion

The results obtained in this study confirm the feasibility of incorporating textile industry waste into bituminous mixtures as a substitute for conventional fibers to mitigate binder drainage while maintaining structural integrity. As established in the introduction, the textile industry is one of the most polluting sectors, generating a high volume of waste that is difficult to reintegrate into its own production cycle. This study demonstrates that recycled cotton textile fibers provide an effective alternative, reinforcing the potential of cross-industry waste utilization to enhance sustainability in road construction.

The inclusion of 0.3% recycled textile fibers in the SMA 11 surf 35/50 mixture proved to be the most effective dosage, offering optimal binder stabilization and mechanical performance. Higher fiber contents, such as 0.6%, resulted in material dispersion issues that negatively affected cohesion and structural integrity The fibers tend to clump together and are not homogeneously distributed within the mixture, resulting in fragile points due to the accumulation of textile fibers. Conversely, mixtures without textile waste (0.0% fibers) exhibited excessive binder drainage, compromising the stability of the mixture and confirming the necessity of fiber reinforcement.

This outcome aligns with prior research highlighting the role of fibers in reducing asphalt binder exudation and improving mechanical stability [13]. Moreover, the results reinforce previous findings that cellulose and textile fibers can effectively replace synthetic stabilizers in SMA mixtures while enhancing performance against mechanical stresses and environmental conditions [15].

The designed mixture met the requirements of Article 544 of the PG-3 [10] in terms of air void content, which was 5.4%, well within the specified 4% to 6% range. The measured maximum density (2.602 g/cm^3) and bulk density (2.462 g/cm^3) confirm that the mixture maintains the necessary compactability for road applications.

Regarding moisture resistance, the water sensitivity test showed an ITSR of 96.30%, exceeding the 90% minimum requirement PG-3 [10]. These results confirm the low susceptibility of the mixture to water-induced damage, ensuring long-term durability. However, the wheel tracking test indicated that while the WTSAIR value of 0.12 mm/1000 cycles for the fiber-reinforced mixture did not fully comply with the $\leq$0.10 mm/1000 cycles limit for T2 traffic, it represents a significant improvement over the 0.32 mm/1000 cycles of the fiber-free mixture. This suggests that slight adjustments in binder type or dosage could enable full compliance with PG-3 standards [10].

In addition to the technical benefits demonstrated through this study, the incorporation of recycled textile fibers into SMA mixtures offers clear economic and environmental advantages. From an economic perspective, the bitumen commonly used in SMA wearing courses has an approximate cost of EUR 650 per tonne [43]. Considering a typical bitumen content of 5.5% by weight for this type of mixture, its contribution to the total cost is estimated at approximately EUR 35.75 per tonne of mixture, representing between 22% and 23% of the overall cost. In contrast, commercial cellulose fibers usually account for 4% to 6% of the total cost, with an average value of EUR 8 per tonne of mixture. Replacing them with recycled textile fibers—particularly when sourced from local waste streams, as in this case study—can significantly reduce the overall cost of the final product.

From an environmental standpoint, this strategy enables the valorization of a waste product that is massively generated in regions such as the Valencian Community, where approximately 150,000 tonnes of textile waste are produced annually [44]. Unlike other materials that can be reintegrated into their original industry, textile waste often lacks effective reintegration routes, leading to accumulation in landfills [16]. Integrating these materials into bituminous mixtures transforms an environmental liability into a valuable engineering resource, aligning with circular economy principles and reducing dependence on virgin raw materials.

Furthermore, sourcing recycled fibers from local textile production hubs minimizes logistical costs and emissions associated with material transport. Since many asphalt plants are located near industrial zones, the incorporation of locally generated textile waste strengthens regional sustainability and fosters synergies between productive sectors. This connection between the textile industry and the road construction sector paves the way for more sustainable solutions, from technical, environmental, and economic perspectives.

The findings of this study confirm that recycled cotton textile fibers can successfully replace traditional stabilizers in SMA mixtures, offering technical, environmental, and economic benefits. The 0.3% fiber dosage was found to be the optimal balance between binder stabilization, mechanical strength, and regulatory compliance. While slight modifications may be required to fully meet the WTS_{AIR} standard for T2 traffic, the results suggest that textile waste incorporation is a viable and scalable strategy for enhancing asphalt mixture performance while contributing to waste reduction and circular economy goals.

The cross-sector utilization of industrial waste offers a sustainable pathway for the pavement industry, demonstrating that materials traditionally deemed non-recyclable in their primary sector can be successfully repurposed to improve infrastructure resilience and environmental sustainability.

Although the results are promising, certain limitations must be acknowledged, and further research is needed to optimize the performance of these mixtures. While the study focused on a fixed bitumen content of 5.3%, adjusting this parameter could improve mechanical resistance and rutting performance, particularly for applications in higher traffic categories. Future studies should investigate the ideal bitumen–fiber balance to maximize performance. The research evaluated binder drainage, water sensitivity, and wheel tracking resistance, but further mechanical tests are needed to assess long-term durability. In particular, tests on fatigue resistance, stiffness modulus, and thermal susceptibility would provide a more comprehensive understanding of the material's behavior under real-world conditions. While laboratory tests provide valuable insights, the next step is to validate these mixtures under real traffic and environmental conditions. Long-term monitoring of trial sections would allow for assessing structural integrity, aging behavior, and maintenance needs over time. Beyond SMA mixtures for surface layers, future research should examine the potential use of textile waste in other asphalt applications, such as base layers, porous asphalt for noise reduction, and warm mix asphalt technologies to enhance sustainability further.

5. Conclusions

This study evaluated the incorporation of recycled cotton textile fibers into SMA mixtures, aiming to enhance mechanical performance and sustainability in road construction.

- These mixtures are suitable for moderate-traffic categories (T2 or lower), especially for secondary roads, urban streets, rural paths, and controlled-traffic industrial areas.
- Textile fibers (0.3%) significantly reduce binder drainage compared to mixtures without fibers.
- High resistance to moisture damage was confirmed (ITSR = 96.3%), ideal for temperate and humid climates.
- The inclusion of textile fibers provides environmental benefits such as waste reduction, lower carbon footprint, and promotion of a circular economy through local resource reuse.
- Wheel tracking tests indicate rutting resistance (0.12 mm/1000 cycles) close to PG-3 limits [10], suggesting potential for further optimization.

Overall, the integration of recycled textile fibers into SMA mixtures is technically feasible and provides substantial environmental and economic benefits, making it a promising sustainable practice for road infrastructure.

Author Contributions: Conceptualization, C.A.-T. and D.L.-C.; methodology, C.A.-T. and D.L.-C.; formal analysis, C.A.-T. and D.L.-C.; investigation, C.A.-T.; data curation, D.L.-C.; writing—original draft preparation, B.O.-C. and C.A.-T.; writing—review and editing, and D.L.-C.; visualization, B.O.-C.; supervision, D.L.-C. All authors have read and agreed to the published version of the manuscript.

Funding: This research was funded by Guerola Áridos y Hormigones, S.L.

Data Availability Statement: The data presented in this study are available on request from the corresponding author due to privacy restrictions.

Conflicts of Interest: The authors declare no conflicts of interest.

References

1. National Academies of Sciences, Engineering, and Medicine. *Recycling and Reclamation of Asphalt Pavements Using in-Place Methods;* The National Academies Press: Washington, DC, USA, 2011.
2. Huang, B.; Li, G.; Vukosavljevic, D.; Shu, X.; Egan, B.K. Laboratory investigation of mixing hot-mix asphalt with reclaimed asphalt pavement. *Transp. Res. Rec.* **2005**, *1929*, 37–45. [CrossRef]
3. Silva, R.V.; de Brito, J.; Dhir, R.K. Properties and composition of recycled aggregates from construction and demolition waste suitable for concrete production. *Constr. Build. Mater.* **2014**, *65*, 201–217. [CrossRef]
4. Llopis-Castelló, D.; Alonso-Troyano, C.; Álvarez-Troncoso, P.; Marzá-Beltrán, A.; García, A. Design of Sustainable Asphalt Mixtures for Bike Lanes Using RAP and Ceramic Waste as Substitutes for Natural Aggregates. *Sustainability* **2022**, *14*, 15777. [CrossRef]
5. Pasetto, M.; Baliello, A.; Giacomello, G.; Pasquini, E. Sustainable solutions for road pavements: A multi-scale characterization of warm mix asphalts containing steel slags. *J. Clean. Prod.* **2017**, *166*, 835–843. [CrossRef]
6. Xu, F.; Zhao, Y.; Li, K. Using waste plastics as asphalt modifier: A review. *Materials* **2021**, *15*, 110. [CrossRef]
7. Mashaan, N.S.; Ali, A.H.; Karim, M.R.; Abdelaziz, M. A review on using crumb rubber in reinforcement of asphalt pavement. *Sci. World J.* **2014**, *2014*, 214612. [CrossRef]
8. Babalghaith, A.M.; Koting, S.; Sulong, N.H.R.; Karim, M.R.; AlMashjary, B.M. Performance evaluation of stone mastic asphalt (SMA) mixtures with palm oil clinker (POC) as fine aggregate replacement. *Constr. Build. Mater.* **2020**, *262*, 120546. [CrossRef]
9. Babalghaith, A.M.; Koting, S.; Ibrahim, M.R.; Mohd, N.S.; Rozali, S.; Muhamad, M.R.; Yusoff, N.I.M. Effect of Black PET Fiber as Additive on the Mechanical Properties of Stone Mastic Asphalt (SMA) Mixtures. In *Advances in Civil Engineering Materials: Selected Articles from the International Conference on Architecture and Civil Engineering (ICACE2020);* Springer: Singapore, 2021; pp. 249–260.
10. Spanish Ministry of Public Works. *Pliego de Prescripciones Técnicas Generales Para Obras de Carreteras Y Puentes de La Dirección General de Carreteras (PG-3);* Ministerio de Fomento, Dirección General de Carreteras; Spanish Ministry of Public Works: Madrid, Spain, 2008.
11. Lugeiyamu, L.; Kunlin, M.; Mensahn, E.S.; Faraz, A. Utilization of waste polyethylene terephthalate (PET) as partial replacement of bitumen in stone mastic asphalt. *Constr. Build. Mater.* **2021**, *309*, 125176. [CrossRef]
12. Ahmadinia, E.; Zargar, M.; Karim, M.R.; Abdelaziz, M.; Shafigh, P. Using waste plastic bottles as additive for stone mastic asphalt. *Mater. Des.* **2011**, *32*, 4844–4849. [CrossRef]
13. Subhy, A.; Lo Presti, D.; Airey, G.; Widyatmoko, I. Rutting analysis of different rubberised stone mastic asphalt mixtures: From binders to mixtures. *Road Mater. Pavement Des.* **2022**, *23*, 2098–2114. [CrossRef]
14. Maurer, D.A.; Malasheskie, G.J. Field performance of fabrics and fibers to retard reflective cracking. *Geotext. Geomembr.* **1989**, *8*, 239–267. [CrossRef]
15. Valdés-Vidal, G.; Calabi-Floody, A.; Mignolet-Garrido, C.; Díaz-Montecinos, C. Effect of a new additive based on textile fibres from end-of-life tyres (ELT) on the mechanical properties of stone mastic asphalt. *Polymers* **2023**, *15*, 1705. [CrossRef]
16. Mokhtari, A.; Nejad, F.M. Mechanistic approach for fiber and polymer modified SMA mixtures. *Constr. Build. Mater.* **2012**, *36*, 381–390. [CrossRef]
17. Yin, J.; Wu, W. Utilization of waste nylon wire in stone matrix asphalt mixtures. *Waste Manag.* **2018**, *78*, 948–954. [CrossRef] [PubMed]
18. Brown, E.R. *Designing Stone Matrix Asphalt Mixtures for Rut-Resistant Pavements;* Transportation Research Board: Washington, DC, USA, 1999; pp. 425–430.
19. Putman, B.J.; Amirkhanian, S.N. Utilization of waste fibers in stone matrix asphalt mixtures. *Resour. Conserv. Recycl.* **2004**, *42*, 265–274. [CrossRef]

20. Behbahani, H.; Nowbakht, S.; Fazaeli, H.; Rahmani, J. Effects of fiber type and content on the rutting performance of stone matrix asphalt. *J. Appl. Sci.* **2009**, *9*, 1980–1984. [CrossRef]
21. Qian, Z.; Lu, Q. Design and laboratory evaluation of small particle porous epoxy asphalt surface mixture for roadway pavements. *Constr. Build. Mater.* **2015**, *77*, 110–116. [CrossRef]
22. Huang, Y.; Liu, Z.; Liu, L.; Zhang, Y.; Xu, Q. Hybrid modification of stone mastic asphalt with cellulose and basalt fiber. *Adv. Mater. Sci. Eng.* **2020**, *2020*, 5671256. [CrossRef]
23. Muniandy, R.; Selim, A.A.; Hassim, S.; Omar, H. Laboratory Evaluation of Ground Tire Rubber in Stone Mastic Asphalt. *J. Eng. Res.* **2004**, *1*, 53–58. [CrossRef]
24. Rodríguez, W.; Rivera, J.; Sevillano, M.; Torres, T. Performance evaluation of Stone Mastic Asphalt (SMA) mixtures with textile waste fibres. *Constr. Build. Mater.* **2024**, *455*, 139125. [CrossRef]
25. Tataranni, P.; Sangiorgi, C. A preliminary laboratory evaluation on the use of shredded cigarette filters as stabilizing fibers for stone mastic asphalts. *Appl. Sci.* **2021**, *11*, 5674. [CrossRef]
26. *UNE-EN 933-2*; Tests for Geometrical Properties of Aggregates—Part 2: Determination of Particle Size Distribution—Test Sieves, Nominal Size of Apertures. Spanish Association for Standardization and Certification (AENOR): Madrid, Spain, 2022.
27. *UNE-EN 1426*; Bitumen and Bituminous Binders. Determination of Needle Penetration. Spanish Association for Standardization and Certification (AENOR): Madrid, Spain, 2016.
28. *UNE-EN 1427*; Bitumen and Bituminous Binders. Determination of the Softening Point. Ring and Ball Method. Spanish Association for Standardization and Certification (AENOR): Madrid, Spain, 2015.
29. *UNE-EN 12591*; Bitumen and Bituminous Binders. Specifications for Paving Grade Bitumens. Spanish Association for Standardization and Certification (AENOR): Madrid, Spain, 2009.
30. *UNE-EN 12593*; Bitumen and Bituminous Binders—Determination of the Fraass Breaking Point. Spanish Association for Standardization and Certification (AENOR): Madrid, Spain, 2015.
31. *UNE-EN 12592*; Bitumen and Bituminous Binders—Determination of Solubility. Spanish Association for Standardization and Certification (AENOR): Madrid, Spain, 2015.
32. *UNE-EN 2592*; Petroleum and Related Products—Determination of Flash and Fire Points. Cleveland Open Cup Method. Spanish Association for Standardization and Certification (AENOR): Madrid, Spain, 2018.
33. *UNE-EN 12607-1*; Bitumen and Bituminous Binders—Determination of the Resistance to Hardening Under Influence of Heat and Air—Part 1: RTFOT Method. Spanish Association for Standardization and Certification (AENOR): Madrid, Spain, 2015.
34. *UNE-EN 12697-18*; Bituminous Mixtures—Test Methods—Part 18: Binder Drainage. Spanish Association for Standardization and Certification (AENOR): Madrid, Spain, 2017.
35. *UNE-EN 12697-6*; Bituminous Mixtures—Test Methods for Hot Mix Asphalt—Part 6: Determination of Bulk Density of Bituminous Specimen. Spanish Association for Standardization and Certification (AENOR): Madrid, Spain, 2020.
36. *UNE-EN 12697-5*; Bituminous Mixtures—Test Methods for Hot Mix Asphalt—Part 5: Determination of the Maximum Density. Spanish Association for Standardization and Certification (AENOR): Madrid, Spain, 2018.
37. *UNE-EN 12697-8*; Bituminous Mixtures—Test Methods for Hot Mix Asphalt—Part 8: Determination of Void Characteristics of Bituminous Specimens. Spanish Association for Standardization and Certification (AENOR): Madrid, Spain, 2018.
38. *UNE-EN 12697-12*; Bituminous Mixtures—Test Methods for Hot Mix Asphalt—Part 12: Determination of the Water Sensitivity of Bituminous Specimens. Spanish Association for Standardization and Certification (AENOR): Madrid, Spain, 2018.
39. *UNE-EN 12697-23*; Bituminous Mixtures—Test Methods for Hot Mix Asphalt—Part 23: Determination of the Indirect Tensile Strength of Bituminous Specimens. Spanish Association for Standardization and Certification (AENOR): Madrid, Spain, 2018.
40. *UNE-EN 12697-30*; Bituminous Mixtures—Test Methods for Hot Mix Asphalt—Part 30: Specimen Preparation by Impact Compactor. Spanish Association for Standardization and Certification (AENOR): Madrid, Spain, 2019.
41. *UNE-EN 12697-22*; Bituminous Mixtures—Test Methods for Hot Mix Asphalt—Part 22: Wheel Tracking. Spanish Association for Standardization and Certification (AENOR): Madrid, Spain, 2020.
42. *UNE-EN 12697-33*; Bituminous Mixtures—Test Methods for Hot Mix Asphalt—Part 33: Specimen Prepared by Roller Compactor. Spanish Association for Standardization and Certification (AENOR): Madrid, Spain, 2019.
43. Ministry of Transport and Sustainable Mobility. *OC 4/2023: Procedure for Price Justification and Support Base 2023*; Government of Spain: Madrid, Spain, 2023.
44. Valencian Institute of Business Competitiveness (IVACE). *ReFashion Project: Textile Recycling and Waste Recovery in the Valencian Community*; Valencian Institute of Business Competitiveness: Valencia, Spain, 2023.

buildings

MDPI

Article

Evaluating Pavement Deterioration Rates Due to Flooding Events Using Explainable AI

Lidan Peng [1,†], Lu Gao [2,*,†], Feng Hong [3,†] and Jingran Sun [4,†]

1 College of Traffic & Transportation, Chongqing Jiaotong University, Chongqing 400074, China; 611240110002@mails.cqjtu.edu.cn
2 Department of Civil and Environmental Engineering, University of Houston, Houston, TX 77204, USA
3 Ingram School of Engineering, Texas State University, San Marcos, TX 78666, USA; fenghong@txstate.edu
4 Center for Transportation Research, The University of Texas at Austin, Austin, TX 78712, USA; jingransun@utexas.edu
* Correspondence: lgao5@central.uh.edu
† These authors contributed equally to this work.

Abstract: Flooding can damage pavement infrastructure significantly, causing both immediate and long-term structural and functional issues. This research investigates how flooding events affect pavement deterioration, specifically focusing on measuring pavement roughness by the International Roughness Index (IRI). To quantify these effects, we utilized 20 years of pavement condition data from TxDOT's PMIS database, which is integrated with flood event data, including duration and spatial extent. Statistical analyses were performed to compare IRI values before and after flooding and to calculate the deterioration rates influenced by flood exposure. Moreover, we applied explainable artificial intelligence (XAI) techniques, such as Shapley Additive Explanations (SHAP) and Local Interpretable Model-Agnostic Explanations (LIME), to assess the impact of flooding on pavement performance. The results demonstrate that flood-affected pavements experience a more rapid increase in roughness compared to non-flooded sections. These findings emphasize the need for proactive flood mitigation strategies, including improved drainage systems, flood-resistant materials, and preventative maintenance, to enhance pavement resilience in vulnerable regions.

Keywords: pavement deterioration; flooding; explainable AI; pavement performance; resilience

1. Introduction

Pavement deterioration due to flooding leads to several types of damage, including reduced pavement strength, increased roughness, rutting, and cracking, which can lead to a loss of pavement service life and increased rehabilitation costs [1]. The economic impact includes direct repair costs, additional user costs related to delays and accidents, and the need for more frequent reconstruction [2]. Structurally, flooding and moisture infiltration weaken pavement layers, reduce the resilient modulus of unbound materials, and cause differential movement, leading to longitudinal cracks and heaves [3]. The severity of deterioration depends on factors such as the pavement structure, drainage, traffic, flood duration, and the condition of the pavement before flooding [4].

To better understand and mitigate these impacts, researchers have employed various methods, including field observations, experimental studies, numerical modeling, and mechanistic–empirical (M-E) simulations. Some studies develop predictive models based

https://doi.org/10.3390/buildings15091452

on experimental measurements of pavement responses before and after flooding [3]. Numerical tools are also employed to simulate pavement responses under various conditions, with the goal of creating algorithms that can generate reasonable results from a limited number of input parameters for network-level applications. M-E design tools such as UPDAPS and TxME are used to analyze pavement performance under flooding by modifying layer moduli to reflect moisture-induced weakening of materials [1]. Machine learning techniques are increasingly used to predict pavement performance after flood events, using historical pavement distress and flood data [5]. Studies have also explored the economic aspects, such as increased maintenance and rehabilitation costs, and proposed frameworks for cost analysis and resource allocation [6]. Furthermore, research has been conducted to define and evaluate pavement resilience, with the aim of developing methods to minimize loss and improve design and construction practices [7].

Despite these advances, assessing flood-induced pavement damage presents several challenges, including the difficulty in obtaining pre- and post-flood data for a comprehensive analysis [8]. Much damage is not visible on the surface immediately after flooding, making it difficult to assess the extent of the problem [9]. Current mechanistic–empirical (ME)-based approaches and climate data modeling may not accurately simulate flooding and extreme precipitation events due to limitations in capturing short-term events and moisture damage [1]. In addition, it is challenging to incorporate all flood-related stressors (flood depth, duration, velocity, debris, and contaminants) into a single model [10]. Factors such as varying pavement section lengths and heterogeneous flood data sources add complexity to network-level assessments [5]. Moreover, accurately predicting long-term deterioration and costs is difficult due to the complex interaction of factors and the variability in agency policies and priorities. Given these limitations, there is a critical need for more in situ data collection and research to improve the accuracy of pavement deterioration assessments and enhance resilience planning [2]. Motivated by these challenges, this research integrates explainable artificial intelligence (XAI) techniques, specifically SHAP and LIME, with advanced machine learning models. This approach not only enhances prediction accuracy but also provides transparent insights into the contributions of key variables.

The objective of this study is to assess flood-induced pavement degradation by analyzing 20 years of TxDOT PMIS data, and measuring changes in the International Roughness Index (IRI) caused by flood events. Various regression models are developed and evaluated, with explainable AI methods, SHAP, and LIME applied to highlight the contributions of key factors such as initial pavement condition, truck traffic, and flood exposure. The study focuses on isolating the impacts of flooding by comparing flooded and non-flooded pavement sections within the studied region, aiming to provide actionable insights for data-driven decision making regarding flood mitigation and maintenance planning.

2. Literature Review

Roads play a crucial role in facilitating transportation as critical infrastructure components. However, they are vulnerable to damage caused by natural disasters, such as floods. Research consistently demonstrates that rainfall negatively impacts pavement conditions over various time periods [11]. Therefore, it is important to understand the effects of floods on roads and pavements and implement appropriate systems and designs to prevent or minimize flood-related damage.

2.1. Impact of Flooding on Pavement Condition and Performance

Flooding significantly compromises pavement structure and functionality, exhibiting various deterioration patterns [12]. The intrusion of water into pavement structures elevates the moisture content within unbound materials, consequently diminishing their modulus and load-bearing capabilities [3]. This saturation process undermines the adhesive and cohesive forces that bind aggregates and asphalt, leading to expedited material degradation, such as asphalt stripping [13]. Moreover, excess moisture in the asphalt layer can result in weakening of aggregate–bitumen bonding and can lead to a higher rate of damage accumulation in the asphalt layer [1]. Therefore, controlling moisture infiltration and ensuring proper drainage are critical in preventing moisture-induced raveling in asphalt pavements [14]. The subsequent reduction in structural support intensifies deflections under load, predisposing the pavement to premature failure mechanisms such as rutting and fatigue cracking. The degree of deterioration is contingent on the attributes of the flood event, the structural composition of the pavement, and prevailing environmental conditions [13]. Furthermore, flood-induced debris can obstruct drainage pathways, exacerbating water retention and prolonging the saturation of pavement layers [1].

The type of damage caused by flooding can be broadly categorized as either visible or hidden [15]. Visible damage includes readily apparent issues such as cave-ins or washouts of the pavement. Hidden damage, on the contrary, encompasses the weakening of the pavement's internal strength and stiffness. While less immediately evident, hidden damage precipitates rapid deterioration and potential catastrophic failures. The specific stressors associated with flooding events include flood depth, duration, velocity, debris, and contaminants [1]. Flood depth and duration contribute to increased saturation within pavement layers, while flood velocity can inflict immediate structural compromise. Debris accumulation can impede drainage efficiency, further extending water exposure. The impact of flooding events demonstrates distinct patterns, including delayed effects, jump effects (characterized by sudden performance decline), a combination of jump and delayed effects, and direct failure scenarios [13].

The existing state of a pavement significantly influences its response to flooding [4]. Pavements initially in good or very good condition experience a sudden decline in quality following a flood. Conversely, pavements already in fair or poor condition do not exhibit a substantial immediate decline but undergo accelerated deterioration compared to non-flooded pavements. Stochastic modeling techniques, such as Monte Carlo simulations within discrete-time Markov chains, can effectively represent these deterioration dynamics. These models facilitate the assessment of pavement condition changes over time under various flooding scenarios, providing valuable insights for infrastructure management.

Effective strategies are available to reduce the adverse effects of flooding on pavement infrastructure. Incorporating efficient drainage systems is critical in mitigating water infiltration and saturation [1]. Employing durable and flood-resistant materials enhances the pavement's capacity to withstand long periods of water exposure. Adopting robust pavement structures reinforces the pavement's capacity to endure flood-induced stresses [12]. Regular inspection, maintenance, and repair protocols are essential for proactively identifying and addressing vulnerabilities. Implementing comprehensive flood management strategies and actively monitoring weather patterns are also crucial for preparedness and responsiveness during flood events [2]. Innovative solutions, such as integrating geotextile-wrapped sand layers, offer enhanced drainage capabilities and effectively mitigate subgrade moisture fluctuations and swelling pressures [15].

2.2. Deterioration Rate Estimation

Estimating deterioration rates resulting from flooding requires a comprehensive analysis of the impact of flood events on the road network's condition over time. This can be a complex task as the rate of deterioration depends on multiple factors, including the intensity and duration of the flooding, the road and infrastructure type, and the quality of maintenance and repair work. One effective approach to estimating road deterioration rates due to flooding involves utilizing historical data on flood events and their influence on the road network. Researchers have conducted different studies and proposed methods for predicting the extent of road degradation during floods and also emphasized the significance of consistent road condition monitoring for timely maintenance and reconstruction efforts.

For instance, Zhang et al. [16] collected data from pavement sections that were submerged due to flooding after Hurricane Katrina and from pavement sections that did not experience flooding. Subsequently, they conducted a systematic analysis and compared the deflection, effective structural number, and resilient modulus of both flooded and non-flooded pavement sections to measure the flood's impact. Chen and Zhang [17] conducted an analysis of pavement performance data before and after Hurricanes Katrina and Rita in Louisiana. The authors calculated and compared the International Roughness Index values and rutting depth before and after flooding, as well as the difference in pavement performance between flooded and non-flooded road sections. Shamsabadi et al. [18] considered the impact of snowstorms and flooding events on pavement performance. They created a regression model incorporating various factors to effectively forecast the increase in IRI. The study highlights the importance of predicting post-natural disaster road conditions for effective road maintenance planning. Sultana et al. [19] examined data from both flooded and non-flooded pavement sections before and after a flooding incident. They observed the flooded section experienced a significant decline in structural strength. This study states the need for long-term monitoring and regular testing of pavements affected by floods to properly assess the impacts of floods on factors such as rutting and roughness. Khan et al. [20] assumed that the IRI value would jump significantly after flooding events and developed a risk-based framework to evaluate the impact of flooding on road conditions. Their findings indicate that well-constructed road with robust structures and high traffic loads, and which is maintained to a high standard, demonstrates great resilience to flooding. Sultana et al. [21] addressed the impact of flooding events on pavement deterioration and emphasized the necessity of developing new deterioration models specifically for flood-affected pavements, and presented two mechanistic–empirical models for rutting and roughness in flood-affected pavements.

Moreover, Vallès-Vallès and Torres-Machi [4] studied the 2013 Colorado floods by using statistical analysis and stochastic Markov chains with Monte Carlo simulations to quantify deterioration rates and service life loss based on the IRI, and their results highlighted accelerated deterioration in flooded pavements. Shariatfar et al. [5] employed a machine learning approach with historical distress data and 2016 Louisiana flood maps to predict post-flooding network-level pavement performance, focusing on key distress types such as roughness and cracking. Hong et al. [22] utilized a GIS analysis and a Markov process to estimate the reduction in network-level pavement performance due to a 100-year flood in Texas over a 10-year horizon, linking the findings to infrastructure resilience and maintenance planning.

2.3. AI Applications in Pavement Deterioration Modeling

AI and deep learning models are being increasingly applied to various aspects of pavement deterioration modeling, offering advancements in detection, prediction, classification, and data management [23,24]. These technologies are crucial for optimizing maintenance planning, extending pavement lifespan, and reducing road maintenance costs [25].

Recurrent neural networks (RNNs), particularly Long Short-Term Memory (LSTM) networks and Gated Recurrent Units (GRUs), are promising in predicting future pavement conditions by analyzing time-series data [26]. These models learn from historical pavement condition data, such as crack severity, rutting depth, and the International Roughness Index (IRI), to forecast future pavement condition indices. Studies have shown that RNN-

LSTM models outperform traditional deep neural networks (DNNs) in pavement damage prediction, achieving lower root-mean-square error (RMSE) values [25]. DNNs are also employed to predict overall evaluation indices like the IRI, crack index (CI), and pavement condition index (PCI) with high accuracy, leveraging data from on-site tests and avoiding deviations from Linear Regression models [27].

In addition to prediction, convolutional neural networks (CNNs) and hybrid models combining CNNs with RNNs (such as CNN-LSTM) are essential for detection and classification tasks. These deep learning models can automatically extract features from raw data, using images captured by vehicle-mounted cameras, unmanned aerial vehicles (UAVs), or satellite imagery to detect pavement distress including cracks, potholes, and other surface deformations [28]. Deep learning techniques often surpass traditional image processing methods in terms of accuracy and efficiency in identifying such damage, as they do not require manual feature extraction [29]. AI models can also detect whether maintenance and reconstruction treatments have been applied to pavement sections by analyzing condition data over time [30]. Moreover, advanced AI techniques such as graph neural networks (GNNs) and convolutional graph neural networks (ConvGNNs) are being explored to improve deterioration modeling accuracy by incorporating the spatial structure of road networks, such as the condition of neighboring sections [31].

2.4. Research Gap

Previous studies have relied on mechanistic–empirical models, numerical simulations, and statistical analyses to examine the impact of flooding on pavement deterioration, but these efforts often face significant limitations. First, many are constrained to a single flood event or a short time frame, restricting their ability to capture long-term deterioration trends or the cumulative effects of multiple flooding incidents across a broader network. These studies often depend on small historical datasets tied to a single flood event, limiting the temporal window and geographic scope, thus reducing the generalizability of the findings to larger networks or diverse flood scenarios. Even studies adopting a network-level perspective or over extended periods tend to focus on probabilistic reductions in performance rather than directly estimating deterioration in flooded pavements, often failing to provide specific insights into deterioration rates or mechanisms affecting individual pavement sections.

To overcome these shortcomings, this study analyzes 20 years of pavement condition data from TxDOT's PMIS database, which covers the entire Texas road network, to provide a comprehensive assessment of flood-induced pavement deterioration across multiple events at a statewide scale. This extensive temporal and geographic scope directly addresses the problems of limited time frames and narrow spatial coverage in prior research. Furthermore, this research integrates explainable AI (XAI) techniques, including SHAP and LIME, into modeling pavement deterioration caused by flooding. While traditional black-box machine learning models offer accurate predictions, they often lack explainability, making it challenging to understand the factors driving pavement degradation. By incorporating XAI, this study enables the identification of key contributors to pavement deterioration at both the aggregate and individual levels.

3. Methodology

In recent years, interpretability has gained increasing importance in machine learning and data science. Complex models such as deep neural networks or ensemble methods (e.g., Random Forest, Gradient-Boosted Trees) have achieved state-of-the-art performance across various domains but often lack transparency. Model-agnostic interpretability methods are designed to address this shortcoming without making assumptions specific to a particular type of model. Two popular methods in this category are Local Interpretable Model-Agnostic Explanations (LIME) [32] and Shapley Additive Explanations (SHAP) [33]. LIME aims to explain the model's predictions by approximating it locally around a specific instance. SHAP, on the other hand, leverages game-theoretic concepts, specifically Shapley values, to attribute feature importance in a more globally consistent manner. In this section, we detail the underlying mathematical foundations of LIME and the Shapley value formulation in SHAP. We also discuss how to interpret the outputs of this methodology in a statistically coherent way.

3.1. LIME

LIME explains the prediction of a complex model f at a specific instance $\mathbf{x}$ by approximating f locally with a simpler and more interpretable model g, often a linear model in the neighborhood of $\mathbf{x}$. The notion of "local" means that LIME focuses on the behavior of f in a small region around the instance $\mathbf{x}$, rather than trying to explain the function f across the entire data distribution [32].

Suppose $\mathbf{x}$ is the instance we wish to explain. LIME samples data points in the vicinity of $\mathbf{x}$ (by perturbing features of $\mathbf{x}$) to form a local neighborhood. Denote this neighborhood by $\mathcal{N}(\mathbf{x})$. For each point $\mathbf{z}$ in $\mathcal{N}(\mathbf{x})$, we can compute

$$\pi_x(\mathbf{z}) = \exp\left(-\frac{D(\mathbf{x},\mathbf{z})^2}{\sigma^2}\right) \tag{1}$$

where $D(\mathbf{x},\mathbf{z})$ is a distance measure (e.g., Euclidean or cosine distance) between the original instance $\mathbf{x}$ and the perturbed instance $\mathbf{z}$. The term $\pi_x(\mathbf{z})$ is often called a weighting kernel, and it assigns higher weights to samples closer to $\mathbf{x}$ [32].

Next, LIME fits a simple interpretable model g (e.g., a linear model) in this local neighborhood by solving

$$\underset{g\in G}{\operatorname{argmin}} \sum_{\mathbf{z}\in\mathcal{N}(\mathbf{x})} \pi_x(\mathbf{z})\big(f(\mathbf{z}) - g(\mathbf{z})\big)^2 + \Omega(g). \tag{2}$$

where
$f(\mathbf{z})$ = the prediction of the original complex model.
$g(\mathbf{z})$ = the prediction of the simple interpretable model (often linear in the feature space of $\mathbf{z}$).
G = the set of all possible interpretable models under consideration (for instance, linear models).
$\Omega(g)$ = a complexity measure for g, which ensures that among all possible local approximations, the simpler ones are favored [32].

In a linear interpretation, if $g(\mathbf{z}) = w_0 + \sum_j w_j z_j$, the coefficients w_j can be interpreted as the contribution of feature j to the prediction. A higher positive w_j indicates a stronger positive contribution of feature j, whereas a negative w_j indicates a negative contribution. The resulting coefficients from the linear model, fitted only to data points near $\mathbf{x}$, provide insight into which features have the greatest impact on the prediction of f in that local region [34].

3.2. SHAP

SHAP is grounded in the concept of Shapley values, which originate from cooperative game theory. In this framework, each feature is regarded as a "player" contributing to the overall prediction of the model. The Shapley value for a given feature quantifies its contribution to the final prediction by considering all possible subsets (coalitions) of other features [33].

Let there be M features in total. For a particular instance $\mathbf{x}$, let the model's output be denoted by $f(\mathbf{x})$. The idea is to distribute $f(\mathbf{x}) - f(\mathbf{0})$ (the difference between the prediction and some baseline model output) among the M features in a fair manner, where $f(\mathbf{0})$ is often chosen as the output of the model under baseline or reference conditions (e.g., when the features are set to average values) [33].

The Shapley value ϕ_j for feature j is given by

$$\phi_j = \sum_{S \subseteq \{1,\ldots,M\} \setminus \{j\}} \frac{|S|!\,(M - |S| - 1)!}{M!} \left[f(S \cup \{j\}) - f(S) \right] \quad (3)$$

where
S = any subset of the set of features not including feature j.
$f(S)$ = the model's output using the subset S of features. In practice, this can be implemented by marginalizing, or in some cases by conditioning on, the other features.
$\frac{|S|!\,(M-|S|-1)!}{M!}$ = a weighting factor that ensures each possible ordering of features is counted appropriately [35].

Conceptually, ϕ_j is computed by looking at the difference in model output when feature j joins every possible coalition S of other features. This is then averaged over all such subsets of features, giving a comprehensive measure of feature j's importance to the final prediction. The Shapley values ϕ_j indicate how much each feature j shifts the model prediction from the baseline $f(\mathbf{0})$ when feature j is introduced into all possible subsets of the other features. A positive Shapley value means that the feature contributes to increasing the prediction relative to the baseline, while a negative value indicates that the feature pulls the prediction below the baseline. Importantly, SHAP values have desirable fairness and consistency properties, making them appealing for interpretability.

4. Case Study

4.1. Data Collection

In this study, data collection and preprocessing is conducted through a sequence of steps, detailed below:

1. Information regarding road closures associated with flooding incidents was gathered from TxDOT. More than 100 significant flooding events were identified and evaluated. Example major flooding events include the post-Ike flooding in Southeast Texas in 2008, the flooding event in Montgomery County in 2014, the Lubbock County flooding incident in 2015, and another major flood occurrence in Montgomery County in 2017.
2. The geographic coordinates of both the beginning and ending points of each affected route were obtained via a Geographic Information System. These coordinates were subsequently cross-referenced with the Texas Department of Transportation's (TxDOT's) road reference system to identify the respective road reference markers defining the start and end of each route.
3. The relevant pavement sections (each section has a length around 0.5 miles) that fell within the range of the identified beginning and ending points were extracted from TxDOT's Pavement Management Information System (PMIS).
4. Pavement sections that had International Roughness Index (IRI) values available prior to and following the year of the respective flooding event were selectively chosen. In the absence of such information, the sections were considered unsuitable for this analysis and consequently discarded from the dataset.

Table 1 shows sample road closures due to flooding collected in this case study. It includes the route names, flood years, start points, end points, and total lengths (in miles) of the affected road sections. These sample road routes include FM0481 in 2014 (from

Eagle Pass to Uvalde, spanning 43.3 miles), FM1908 in 2014 (from Spofford to Quemado, covering 20.6 miles), and SH0131 in 2014 (from Eagle Pass to Brackettville, with a length of 33.1 miles). Figure 1 provides a visual representation of these sample locations within the study area.

Table 1. Sample flooded road routes.

Route Name	Flood Year	Start Point	End Point
FM0481	2014	Eagle Pass	Uvalde
FM1908	2014	Spofford	Quemado
SH0131	2014	Eagle Pass	Brackettville
FM2854	2017	Loop 336	Pinewood Dr.
US0087	2015	FM 41 south of Lubbock	Lamesa
FM0597	2015	Cochran county line	FM 303
FM0179	2015	Lynn county line	US 87 in Lamesa
US0180	2015	FM 829	SH 137
FM0366	2008	Intersection with FM 347	N.A.
BU0090Y	2008	FM 1006	Interstate 10

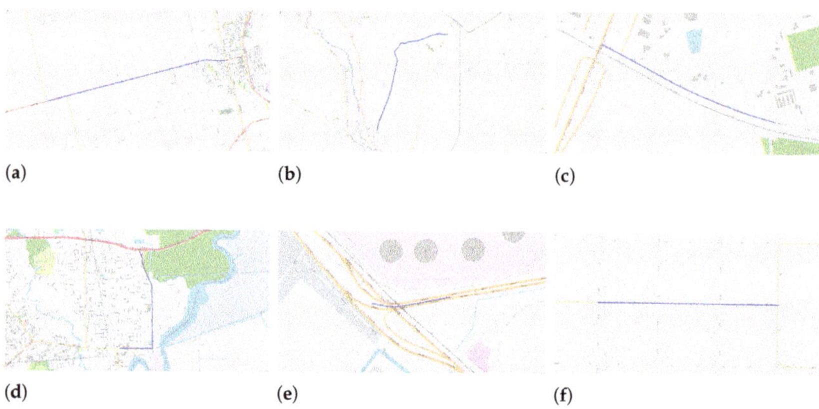

Figure 1. Locations of sample routes. (**a**) US0180, FM0179, and US0087; (**b**) FM1908, SH0131, and FM0481; (**c**) FM2854; (**d**) BU0090Y; (**e**) FM0366; (**f**) FM0597.

4.2. Data Preprocessing

The TxDOT PMIS is a dataset containing a comprehensive pavement condition inventory that includes various pavement performance indicators such as potholes, cracking, and rutting. However, not all features are available for every road segment. The IRI is the only condition indicator available for all road segments and is used as the primary indicator in this case study. Moreover, we selected other variables that represent traffic and environmental factors to provide a more comprehensive analysis. The key features included in the dataset are as follows:

- TX_CONDITION_SCORE: Pavement condition score (0–100). 100 represents the best score and 0 represents the worst score.
- TX_DISTRESS_SCORE: Pavement distress score (0–100).
- TX_IRI_AVERAGE_SCORE: International Roughness Index (IRI) average score.
- TX_TRUCK_AADT_PCT: Percentage of daily traffic that is trucks.
- TX_CURRENT_18KIP_MEAS: A measurement related to 18-KIP equivalent single-axle loads.

- TX_PVMNT_TYPE_DTL_RD_LIFE_CODE: This variable represents the predominant travel lane pavement type during the data collection year for a given section, derived from RLS pavement layer information. It is stored as a string with a maximum length of 100 characters. The possible values include 01—continuously reinforced concrete (CRCP), 02—jointed reinforced concrete (JRCP), 03—jointed plain concrete (JPCP), 04—thick asphaltic concrete (greater than 5.5″), 05—medium-thickness asphaltic concrete (2.5–5.5″), 06—thin asphaltic concrete (less than 2.5″), 07—composite (asphalt-surfaced concrete or ACP on top of a heavily stabilized base), 08—widened composite pavement, and 09—overlaid and widened asphaltic concrete pavement.
- CLIMATE_ZONES: Climate zones were defined based on temperature and precipitation data obtained from the National Oceanic and Atmospheric Administration (NOAA) database. A 30-year annual average of temperature and precipitation was used to represent the climate. For counties without a weather station, data from adjacent counties were averaged, while for counties with multiple weather stations, the average of those stations was used. Counties were grouped into climate zones corresponding to the west, east, north, south, and central regions.
- TX_RURAL_URBAN_CODE: Categorical code indicating if a road section is rural or urban.
- Flood: Binary indicator (1 if the road segment was recorded as flooded that year, 0 otherwise).

The target variable is next year's TX_IRI_AVERAGE_SCORE. A total of 10,022 data records were collected for this case study. Each data point represents a pavement section of length around 0.5 miles. Table 2 summarizes the descriptive statistics of the numerical variables. This includes the mean, standard deviation, minimum, maximum, and quartiles. Among the 10,022 pavement sections, 501 (5%) were classified as flooded, while the remaining 9521 (95%) were identified as non-flooded. The data points were selected based on the availability of key features. Pavement sections with missing values for the features listed in Table 2 were excluded from the statistical analysis.

Table 2. Descriptive statistics of the numerical features.

Feature	Mean	Std. Dev.	Min	25%	Max
TX_CONDITION_SCORE	93.91	13.87	0.00	97.00	100.00
TX_DISTRESS_SCORE	95.70	11.35	0.00	99.00	100.00
TX_IRI_AVERAGE_SCORE	100.61	54.17	26.00	57.00	314.00
TX_TRUCK_AADT_PCT	17.60	8.52	0.00	13.10	56.90
TX_CURRENT_18KIP_MEAS	1096.57	978.45	0.00	200.00	8123.00
TX_PVMNT_TYPE_DTL_RD_LIFE_CODE	8.74	1.98	1.00	6.00	10.00
CLIMATE_ZONES_encoded	1.83	0.98	0.00	2.00	3.00
TX_RURAL_URBAN_CODE	1.03	0.21	1.00	1.00	4.00
Flood	0.05	0.21	0.00	0.00	1.00

Figure 2 presents the Pearson correlation heatmap for features used in this case study. A high positive correlation (0.87) is observed between TX_CONDITION_SCORE and TX_DISTRESS_SCORE, which indicates that the overall pavement condition rating is strongly influenced by the distress score. TX_CONDITION_SCORE shows a moderate negative correlation (−0.43) with TX_IRI_AVERAGE_SCORE. The negative relationship arises because the IRI value indicates worse pavement roughness as it increases, whereas a higher pavement condition score indicates better pavement condition. Traffic-related variables, such as TX_TRUCK_AADT_PCT, show weak correlations with most pavement condition indicators, with the highest correlation (0.22) observed with TX_IRI_AVERAGE_SCORE. The correlation between Flood and current pavement condition variables is relatively weak.

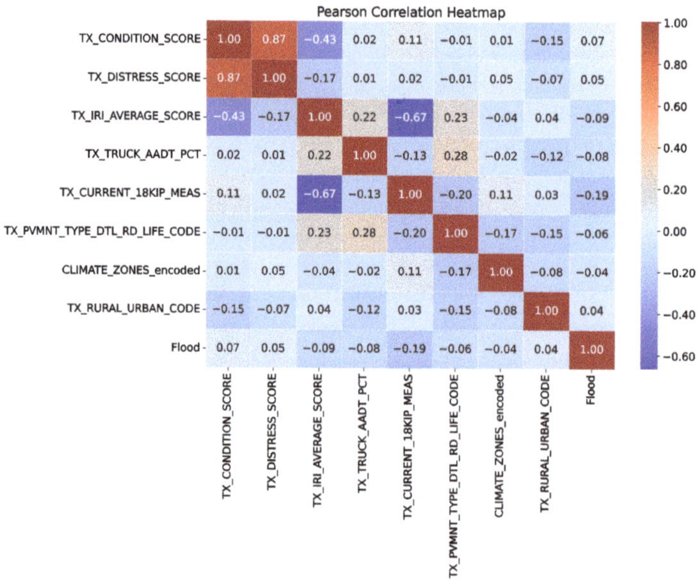

Figure 2. Correlation matrix of selected numerical features.

4.3. Statistical Analysis

4.3.1. IRI Comparison of Before and After Flooding

Figure 3 illustrates the changes in IRI values, measured in inches per mile (in/mi), across multiple roadway routes from before to after flooding events. Each subplot corresponds to a specific route (e.g., FMD481, SH0131, US0087) and displays pre-flood (one year prior to flooding) and post-flood (one year after flooding) IRI measurements for individual sections within the route. For instance, route FMD481 (pre-flood 2013, post-flood 2015) shows the post-flood IRI value of one section reaching up to 200 in/mi, significantly higher than the pre-flood level. Similar trends are evident in other routes, such as US0087 and FM0179, where post-flood IRI values rise sharply.

While the overall trend confirms higher (worse) post-flood IRI values, variability exists among sections. For example, SH0131 (pre-flood 2013, post-flood 2015) exhibits minimal deterioration in some sections. This inconsistency may result from localized flood impacts, temporary route closures by TxDOT during floods, or post-flood maintenance interventions that restored certain segments.

To ensure data reliability, sections lacking complete IRI records during the two-year analysis window (one year before and after flood) were excluded. Also, sections with improved IRI values (potentially due to maintenance) were omitted to isolate flood-induced deterioration. In pavement performance modeling, when maintenance records are unavailable it is common to exclude pavement condition data that show improvement over consecutive years [36]. This is because pavement conditions are not expected to improve without maintenance intervention. In this case study, maintenance records for the analyzed road segments are unavailable. Therefore, if IRI values of a road segment decrease after flooding, it is assumed that maintenance was performed, and these data points are excluded from the statistical analysis.

The analysis reveals an average IRI increase of 8.46 in/mi over two years, with a high standard deviation (10.53 in/mi), indicating significant variability in deterioration severity across sections. This underscores that while flooding generally exacerbates pavement roughness, the extent of damage is highly dependent on localized factors such as flood intensity, road segment vulnerability, and maintenance responsiveness.

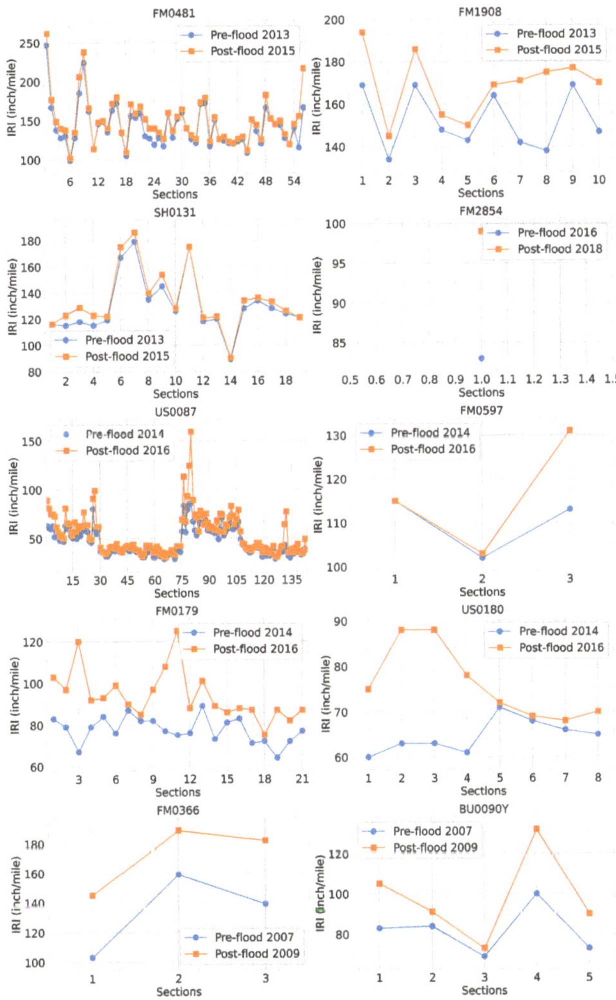

Figure 3. IRI before and after the flooding events.

4.3.2. IRI Deterioration Rate: Comparison Before and After Flooding

Figure 4 illustrates the IRI values during three different years: three years prior to flooding, one year before flooding, and one year after flooding. The chosen interval of two years allows for a comparison of the deterioration rate before and after the flooding event. One-year intervals were not used due to the lack of precise data collection dates for the IRI values. Consequently, it is difficult to determine whether the IRI at the time of flooding was measured before or after the event occurred. The IRI values presented in Figure 4 represent the average values across sections in the same route. When calculating the average, sections lacking IRI data in these three years were excluded. Furthermore, sections where the IRI values improved, indicating potential maintenance effects, were also excluded. For this reason, not all routes were plotted in the figure. The figure shows that, in general, the IRI deterioration rate increases after flooding. However, the rate of deterioration varies across different segments. Some sections experience a much sharper decline in condition, as seen in certain data points for US0087, where roughness increases significantly after flooding. In contrast, there are cases like SH0131, where the impact of flooding appears minimal, with IRI values remaining relatively stable.

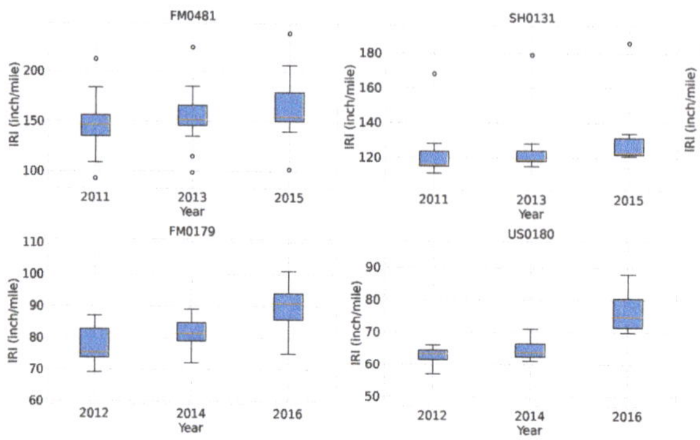

Figure 4. IRI before and after flooding events.

4.3.3. Flooded and Non-Flooded Section Comparison

Figure 5 shows a comparison of the IRI deterioration rate between sections affected by flooding and sections outside the flooding region nearby. In order to compare them under the same condition, both flooded and non-flooded sections are collected from the same route. The x-axis represents different sections and the y-axis represents the IRI difference between the year after flooding and the year before flooding, representing the IRI deterioration in two years. As can be seen from Figure 5, the differences between the flooded and non-flooded sections range from 0.3 to 15 in/mi. The average IRI change difference between flooded and non-flooded sections is 5.6 in/mi.

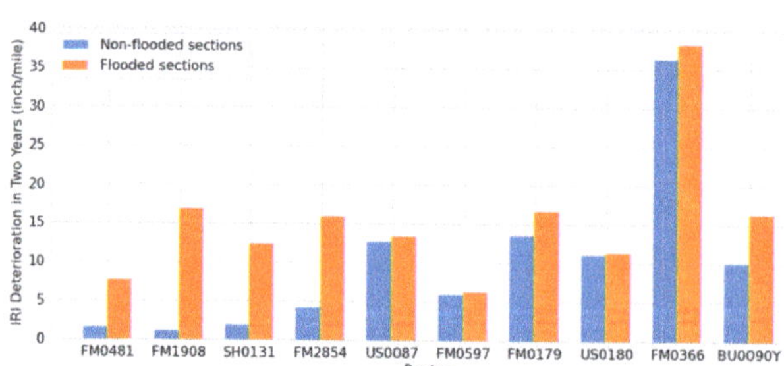

Figure 5. IRI deterioration rates: Flooded vs. non-flooded sections.

4.4. XAI Modeling

4.4.1. Machine Learning Models

A total of six regression models were trained and evaluated to predict next year's IRI: Linear regression (LR), Ridge Regression (Ridge), Lasso Regression (Lasso), Decision Tree (DT), Random Forest (RF), and Gradient Boosting Regression (GBR). The dataset was divided into training (80%) and test (20%) subsets through random sampling to ensure a balanced representation of pavement conditions and external factors.

We implemented all models using the scikit-learn package and initially used default settings. Hyperparameter tuning was performed to optimize performance for certain models. For the Ridge and Lasso models, we adjusted the regularization strength (α) using a grid search over values from 10^{-3} to 10^2. The DT model was optimized by tuning the maximum tree depth (2–20), the minimum samples required to split a node (2–10), and the minimum samples per leaf (1–5). RF tuning included adjusting the number of trees (50, 100, 200), maximum depth (5–20), and minimum samples per split (2, 5, 10), along with feature subsampling. GBR was optimized by adjusting the learning rate (0.001, 0.01, 0.1), number of estimators (100, 300, 500), maximum tree depth (3, 5, 10), and subsampling ratios (0.5, 0.75, 1.0). The best hyperparameters were determined using 5-fold cross-validation on the training set, selecting the combination that minimized mean squared error (MSE).

Each of the six regression models was evaluated using mean squared error (MSE), mean absolute error (MAE), and R^2 score on the test set. Table 3 summarizes the performance. Among the models evaluated, GBR demonstrated the best performance, achieving the lowest mean squared error (MSE) of 247.0450, the lowest mean absolute error (MAE) of 8.6095, and the highest R^2 score of 0.9310. This result suggests that GBR effectively captures complex relationships within the dataset. The LR, Ridge, and Lasso exhibited nearly identical performance, with MSE values around 263 and R^2 scores of approximately 0.9263–0.9266. Lasso Regression achieved a slightly better MSE (262.8292) and R^2 score (0.9266) compared to the standard Linear Regression model, though the improvement was marginal. The DT model performed significantly worse than the other models, with the highest MSE (360.7363) and MAE (10.6944), resulting in a notably lower R^2 score of 0.8993. The RF model achieved slightly better performance than the DT, with an MSE of 264.0790 and an R^2 score of 0.9262. Since GBR outperformed all other models in terms of predictive accuracy, we use it for further XAI analysis to better understand the contributions of different features to pavement deterioration.

Table 3. Comparison of regression models for predicting next year's IRI.

Model	MSE	MAE	R^2 Score
Linear Regression (LR)	263.8513	8.8476	0.9263
Ridge Regression (Ridge)	263.7546	8.8463	0.9263
Lasso Regression (Lasso)	262.8292	8.8517	0.9266
Decision Tree (DT)	360.7363	10.6944	0.8993
Random Forest (RF)	264.0790	8.5476	0.9262
Gradient Boosting (GBR)	247.0450	8.6095	0.9310

4.4.2. Feature Importance via SHAP

SHAP values were computed to assess the global contribution of each feature to the model's predictions. Figure 6 shows a summary plot of the SHAP values for all features. The SHAP summary plot provides insights into how key variables influence the predictions of a GBR model, which is trained to predict next year's IRI. The features are ranked by their mean absolute SHAP values, with the most influential variables appearing at the top. The results indicate that TX_IRI_AVERAGE_SCORE is the most critical predictor, followed by CLIMATE_ZONES, TX_TRUCK_AADT_PCT, and TX_CURRENT_18KIP_MEAS.

The SHAP values along the x-axis indicate how each feature affects the model's predictions. Positive SHAP values increase the predicted IRI, meaning worse pavement conditions, while negative SHAP values decrease the predicted IRI, indicating better pavement performance. The color coding in the SHAP plot represents the feature values, where red denotes high values and blue denotes low values.

TX_IRI_AVERAGE_SCORE has the greatest impact on next year's IRI value. Higher current IRI scores (red) significantly increase the predicted IRI, suggesting that roads that are already rough are likely to become even rougher next year. Lower current IRI scores (blue) tend to have a neutral or slightly negative impact, indicating that smoother roads are likely to remain relatively smooth or improve slightly. TX_TRUCK_AADT_PCT has a moderate impact. A higher percentage of truck traffic (red) increases the predicted IRI, indicating that roads with more truck traffic are likely to become rougher due to the heavy loads accelerating pavement deterioration. A lower percentage (blue) has a neutral or slightly negative impact, suggesting that roads with less truck traffic are likely to experience less increase in roughness or maintain smoother surfaces. For TX_CURRENT_18KIP_MEAS, the balanced distribution and moderate spread of SHAP values indicate that its influence varies depending on the specific load magnitude across data instances. Low values of TX_CURRENT_18KIP_MEAS (blue) are associated with higher SHAP values, indicating that weaker pavements deteriorate faster.

The Flood feature also shows a tendency toward positive SHAP values, suggesting that flood-prone areas experience higher predicted IRI values, indicating accelerated pavement deterioration. This outcome aligns with the physical effects of flooding on roadways. The SHAP distribution of the Flood variable indicates that flood-exposed roadways are more likely to suffer from worsening pavement conditions in the following year. The effect of flooding is evident, as instances with high flood exposure show increased SHAP values, indicating the role of extreme weather events in pavement degradation.

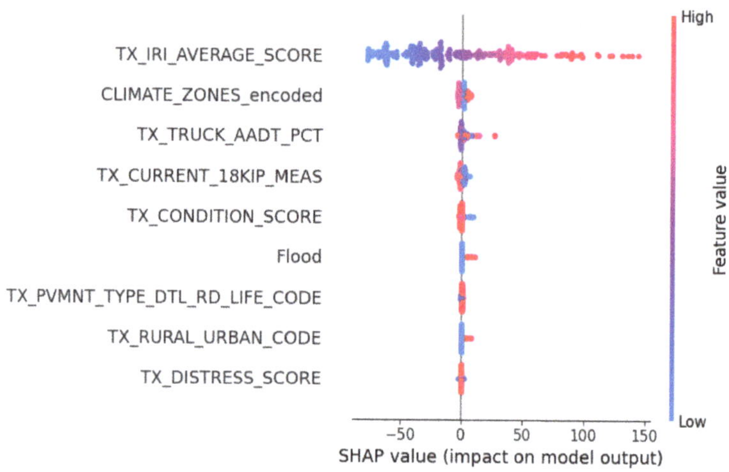

Figure 6. SHAP summary plot for the gradient boosting model.

4.4.3. Local Interpretations via LIME

Figure 7 presents a LIME analysis for one pavement segment and the impact of individual features on the predicted IRI. The bars represent the direction and magnitude of each feature's contribution to the model's prediction, where green bars indicate a positive impact (leading to a higher predicted IRI, meaning worse pavement condition) and red bars indicate a negative impact (leading to a lower predicted IRI, meaning better pavement condition). The most influential feature in this instance is TX_IRI_AVERAGE_SCORE, which falls within the range of 90.00 to 100.00. Its green bar extending farthest to the right indicates that a high initial roughness significantly contributes to a higher predicted IRI value. The Flood feature, with a condition of Flood > 0.0, appears with a green bar positioned around +5 on the x-axis, indicating that flood occurrence contributes to an increase in IRI by approximately 5 units. This aligns with the expectation that flooding accelerates pavement deterioration. The moderate impact suggests that while flooding plays a role in roughness progression, other factors such as initial pavement condition may have a greater influence in this particular instance. TX_CONDITION_SCORE has a negative impact on IRI, which is expected since higher condition scores indicate better pavement quality and are associated with lower roughness. Similarly, TX_DISTRESS_SCORE follows the same trend: higher distress scores in this range correlate with lower IRI values. Moreover, TX_CURRENT_18KIP_MEAS has a minor negative impact, indicating that higher axle loads are linked to reduced roughness in this particular case. This could be due to the presence of more robust pavement structures on roads designed to accommodate heavy traffic, leading to better surface conditions. TX_TRUCK_AADT_PCT, representing truck traffic percentage, has a red bar, implying that in this specific case, the given level of truck traffic is associated with a lower predicted IRI. This may be due to well-maintained roads in high-traffic areas or the presence of more durable pavement designs. Overall, Figure 7 provides a localized interpretation of how various factors influence the IRI prediction for one specific pavement segment. It confirms that flood exposure significantly contributes to pavement deterioration, while some features, such as truck traffic percentage, may correlate with reduced roughness in this specific case.

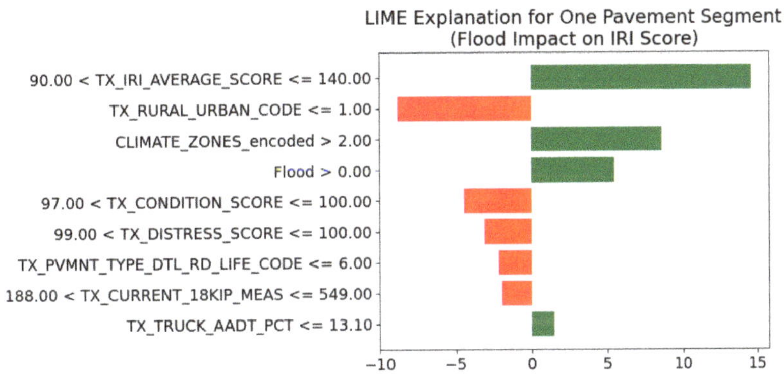

Figure 7. Example LIME explanation for one pavement segment, indicating that flood occurrence contributes to higher IRI (indicating worse condition). Green bars indicate positive contributions to IRI; red bars indicate negative contributions.

4.5. Discussion

This research provides important insights into the impact of flooding on pavement deterioration. The analysis reveals an average IRI increase of 8.46 in/mi over two years, with a high standard deviation of 10.53 in/mi, indicating significant variability in deterioration severity across sections. While flooding generally exacerbates pavement roughness, the extent of damage largely depends on localized factors such as flood intensity, road segment vulnerability, and maintenance responsiveness. On average, the deterioration rate of IRI increased compared to the same two-year period before the flooding, which highlights the substantial impact of flood events. Moreover, the difference in IRI changes between flooded and non-flooded sections ranges from 0.3 to 15 in/mi, with an average difference of 5.6 in/mi. These findings align with previous studies, such as Shariatfar et al. [5], who observed a slightly higher rate of IRI increase in flooded areas after a flood. Similarly, Vallès-Vallès and Torres-Machi [4] reported a sudden drop in the Riding Index (RIDX) of 27 and 11 points for pavements in very good and good condition, respectively, corresponding to a rise in IRI and a quantified service life reduction of 14.3% for pavements in good condition, suggesting an accelerated deterioration rate due to flooding.

The application of SHAP analysis highlights that flooding, along with truck traffic percentage and initial pavement condition, is among the most influential factor contributing to pavement degradation. Moreover, LIME explanations confirm that flood exposure significantly contributes to increased roughness. The integration of XAI techniques into pavement deterioration analysis not only enhances model transparency but also facilitates data-driven decision making for transportation agencies. SHAP analysis allows for a global understanding of feature importance. On the other hand, LIME provides local interpretability, helping identify specific instances where unexpected deterioration trends occur. The combination of these techniques can bridge the gap between traditional black-box machine learning models and actionable insights.

The findings from this study can guide transportation agencies in prioritizing maintenance for flood-prone roads by focusing on sections with higher IRI increases and variability. Proactive scheduling of preventive maintenance, such as resurfacing and drainage improvements, can help extend the service life of pavement and prevent severe deterioration. Moreover, applying flood-resistant materials and improved drainage systems in high-risk areas can enhance infrastructure resilience.

5. Conclusions

This study evaluates the impact of flooding on pavement deterioration using 20 years of pavement condition data from TxDOT's PMIS database, focusing on roughness changes measured by the IRI. The findings indicate that flood-affected pavement sections experience significantly faster deterioration than non-flooded roads. On average, the IRI increased by 8.46 in/mi over two years for flooded sections. Moreover, the deterioration rate after flooding was significantly higher than the rate observed in the two years before flooding. When comparing flooded and non-flooded pavement sections, the study found that flooded roads deteriorated by an additional 5.6 in/mi on average, with some sections even experiencing an extreme deterioration difference of up to 15 in/mi. To understand the underlying causes of pavement degradation, this research applied explainable AI (XAI) techniques, including SHAP and LIME. The results identified initial pavement roughness (IRI values), truck traffic percentage, flood exposure, and climate zone conditions as the most influential factors affecting pavement deterioration. These findings will help transportation agencies focus maintenance efforts on flood-prone roads with high IRI increases and deterioration variability.

This research has some limitations that should be acknowledged. First, the study is based on a limited number of flooding events in the case study, which may restrict the generalization of our findings. To enhance the robustness of our conclusions, it would be beneficial to include a more extensive range of flooding events with more diverse geographic locations in future studies. Second, the availability of data points posed a constraint on our research. As we only had access to PMIS data up to 2018, some recent flooding events had to be excluded from this study. Moving forward, efforts will be made to address these two limitations and expand the scope of the research to encompass a broader set of events and more up-to-date data. Third, maintenance activities that might have influenced post-flood IRI values could not be entirely accounted for, potentially affecting the precision of the deterioration estimates. Future research should aim to incorporate past maintenance work records. Fourth, while various pavement performance indicators could provide additional insights into the impact of flooding, these parameters (e.g., cracking, rutting, and raveling) were either not available in the dataset or were too incomplete to allow for statistical analysis in this study. Although data on potholes, cracking, and rutting were present, they covered only a small fraction of the analyzed sections. In contrast, IRI values were available for all sections, which is the reason why IRI was selected as the primary indicator. Future studies should explore datasets with more comprehensive pavement condition metrics to enhance the analysis.

Author Contributions: Conceptualization, L.G.; methodology, L.P.; software, L.G.; validation, L.P., L.G. and F.H.; formal analysis, F.H.; investigation, L.P.; resources, J.S.; data curation, L.P.; writing—original draft preparation, L.P.; writing—review and editing, L.G. and J.S.; visualization, L.P.; supervision, L.G.; project administration, L.G. All authors have read and agreed to the published version of the manuscript.

Funding: This research received no external funding.

Data Availability Statement: The data presented in this study are available on request from the corresponding author due to legal reasons.

Conflicts of Interest: The authors declare no conflicts of interest.

References

1. Abdollahi, S.F.; Kutay, M.E.; Lanotte, M. UPDAPS-Flood: A mechanistic-empirical flexible pavement analysis tool to evaluate the effect of flooding events on flexible pavement performance. *Int. J. Pavement Eng.* **2024**, *25*, 2387743. [CrossRef]
2. Hong, F.; Prozzi, J. Assessment of flooding impact on thin pavement structure in Texas coastal region. *Int. J. Transp. Sci. Technol.* 2024, *in press*.
3. Asadi, M.; Nazarian, S.; Mallick, R.B.; Tirado, C. Computational process for quantifying the impact of flooding on remaining life of flexible pavement structures. *J. Transp. Eng. Part B Pavements* **2020**, *146*, 04020060. [CrossRef]
4. Vallès-Vallès, D.; Torres-Machi, C. Deterioration of flexible pavements induced by flooding: Case study using stochastic Monte Carlo simulations in discrete-time Markov chains. *J. Infrastruct. Syst.* **2023**, *29*, 05022009. [CrossRef]
5. Shariatfar, M.; Lee, Y.C.; Choi, K.; Kim, M. Effects of flooding on pavement performance: A machine learning-based network-level assessment. *Sustain. Resilient Infrastruct.* **2022**, *7*, 695–714. [CrossRef]
6. Matini, N.; Qiao, Y.; Sias, J.E. Development of time–depth–damage functions for flooded flexible pavements. *J. Transp. Eng. Part B Pavements* **2022**, *148*, 04022011. [CrossRef]
7. Nivedya, M.; Tao, M.; Mallick, R.B.; Daniel, J.S.; Jacobs, J.M. A framework for the assessment of contribution of base layer performance towards resilience of flexible pavement to flooding. *Int. J. Pavement Eng.* **2020**, *21*, 1223–1234. [CrossRef]
8. Yu-Shan, A.; Shakiba, M. Flooded pavement: Numerical investigation of saturation effects on asphalt pavement structures. *J. Transp. Eng. Part B Pavements* **2021**, *147*, 04021025. [CrossRef]
9. Mallick, R.B.; Tao, M.; Daniel, J.S.; Jacobs, J.M.; Veeraragavan, A. Combined model framework for asphalt pavement condition determination after flooding. *Transp. Res. Rec.* **2017**, *2639*, 64–72. [CrossRef]
10. Achebe, J.; Oyediji, O.; Saari, R.K.; Tighe, S.; Nasir, F. Incorporating flood hazards into pavement sustainability assessment. *Transp. Res. Rec.* **2021**, *2675*, 1025–1042. [CrossRef]
11. Sultana, M.; Chai, G.; Martin, T.; Chowdhury, S. Modeling the postflood short-term behavior of flexible pavements. *J. Transp. Eng.* **2016**, *142*, 04016042. [CrossRef]
12. Chen, X.; Wang, H. Evaluation of pavement resilience to flooding with inverted pavement structure. *Road Mater. Pavement Des.* **2024**, *25*, 2709–2728. [CrossRef]
13. Lu, D.; Tighe, S.L.; Xie, W.C. Impact of flood hazards on pavement performance. *Int. J. Pavement Eng.* **2020**, *21*, 746–752. [CrossRef]
14. Do, T.C. Engineering properties-based parameters used for moisture damage evaluation of asphalt mixtures: A review. *Can. J. Civ. Eng.* **2023**, *51*, 390–398. [CrossRef]
15. Chen, Q.; Zhang, Z.; Gaspard, K. Case study on the impact of flooding and inundation on pavement performance. *Transp. Res. Rec.* **2023**, *2677*, 157–168. [CrossRef]
16. Zhang, Z.; Wu, Z.; Martinez, M.; Gaspard, K. Pavement structures damage caused by Hurricane Katrina flooding. *J. Geotech. Geoenviron. Eng.* **2008**, *134*, 633–643. [CrossRef]
17. Chen, X.; Zhang, Z. Effects of Hurricanes Katrina and Rita flooding on Louisiana pavement performance. In *Pavement Materials, Structures, and Performance*; ASCE Library: Reston, VA, USA, 2014; pp. 212–221.
18. Shamsabadi, S.S.; Tari, Y.S.H.; Birken, R.; Wang, M. Deterioration forecasting in flexible pavements due to floods and snow storms. In Proceedings of the EWSHM—7th European Workshop on Structural Health Monitoring, Nantes, France, 8–11 July 2014.
19. Sultana, M.; Chai, G.; Martin, T.; Chowdhury, S. A study on the flood affected flexible pavements in Australia. In Proceedings of the 9th International Conference on Road and Airfield Pavement Technology, Dalian, China, 9–13 August 2015; pp. 9–13.
20. Khan, M.U.; Mesbah, M.; Ferreira, L.; Williams, D.J. Assessment of flood risk to performance of highway pavements. In *Institution of Civil Engineers-Transport*; Thomas Telford Ltd.: London, UK, 2017; Volume 170, pp. 363–372.
21. Sultana, M.; Chai, G.; Chowdhury, S.; Martin, T.; Anissimov, Y.; Rahman, A. Rutting and roughness of flood-affected pavements: Literature review and deterioration models. *J. Infrastruct. Syst.* **2018**, *24*, 04018006. [CrossRef]

22. Hong, F.; Wu, H.; Li, J.; Paul, E. Evaluation of flood-vulnerable pavement network in support of resilience in pavement system management. *Transp. Res. Rec.* **2023**, *2677*, 474–482. [CrossRef]
23. Choi, S.; Do, M. Development of the road pavement deterioration model based on the deep learning method. *Electronics* **2019**, *9*, 3. [CrossRef]
24. Lebaku, P.K.R.; Gao, L.; Lu, P.; Sun, J. Deep Learning for Pavement Condition Evaluation Using Satellite Imagery. *Infrastructures* **2024**, *9*, 155. [CrossRef]
25. Lee, Y.; Sun, J.; Lee, M. Development of deep learning based deterioration prediction model for the maintenance planning of highway pavement. *Korean J. Constr. Eng. Manag.* **2019**, *20*, 34–43.
26. Hosseini, S.A.; Alhasan, A.; Smadi, O. Use of deep learning to study modeling deterioration of pavements a case study in Iowa. *Infrastructures* **2020**, *5*, 95. [CrossRef]
27. Guo, X.; Wang, N.; Li, Y. Enhancing pavement maintenance: A deep learning model for accurate prediction and early detection of pavement structural damage. *Constr. Build. Mater.* **2023**, *409*, 133970. [CrossRef]
28. Chen, X.; Zhang, X.; Li, J.; Ren, M.; Zhou, B. A new method for automated monitoring of road pavement aging conditions based on recurrent neural network. *IEEE Trans. Intell. Transp. Syst.* **2022**, *23*, 24510–24523. [CrossRef]
29. Inácio, D.; Oliveira, H.; Oliveira, P.; Correia, P. A low-cost deep learning system to characterize asphalt surface deterioration. *Remote Sens.* **2023**, *15*, 1701. [CrossRef]
30. Gao, L.; Han, Z.; Chen, Y. Deep learning–based pavement performance modeling using multiple distress indicators and road work history. *J. Transp. Eng. Part B Pavements* **2023**, *149*, 04022061. [CrossRef]
31. Gao, L.; Yu, K.; Lu, P. Considering the spatial structure of the road network in pavement deterioration modeling. *Transp. Res. Rec.* **2024**, *2678*, 153–161. [CrossRef]
32. Ribeiro, M.T.; Singh, S.; Guestrin, C. "Why should i trust you?" Explaining the predictions of any classifier. In Proceedings of the 22nd ACM SIGKDD International Conference on Knowledge Discovery and Data Mining, San Francisco, CA, USA, 13–17 August 2016; pp. 1135–1144.
33. Lundberg, S. A unified approach to interpreting model predictions. *arXiv* **2017**, arXiv:1705.07874.
34. Molnar, C. *Interpretable Machine Learning*; Lulu. com: Morrisville, NC, USA, 2020.
35. Shapley, L.S. A Value for n-Person Games. 1953. Available online: https://www.rand.org/content/dam/rand/pubs/papers/2021/P295.pdf (accessed on 5 February 2025).
36. Mishalani, R.G.; Madanat, S.M. Computation of infrastructure transition probabilities using stochastic duration models. *J. Infrastruct. Syst.* **2002**, *8*, 139–148. [CrossRef]

buildings

MDPI

Article

Development and Application of an AI-Based Automatic Identification System for Rural Road Distress and Maintenance Management

Longjin Chen, Hang Zhang *, Dongfang Li, Yulong Li, Jiang Lou and Kenxuan Fu

Zhejiang Highway Technicians College, Hangzhou 310023, China; chenlongjin088@yeah.net (L.C.); lidongfang2018@yeah.net (D.L.); liyulongsw@163.com (Y.L.); zjgl_loujiang@163.com (J.L.); fukenxuan0202@163.com (K.F.)
* Correspondence: zhanghang0108@yeah.net

Abstract: With the continuous expansion of rural road construction and increasing management demands, traditional rural road inspection and maintenance models are becoming insufficient to meet current needs. The analysis of inspection results and the development of maintenance plans are often delayed. To address these challenges, this paper proposes a rural road distress sample recognition and annotation method based on machine vision techniques, and establishes a corresponding disease target identification sample database. The method is trained and validated using the U-Net algorithm, achieving an accuracy of 94.95%. Additionally, a lightweight detection system is developed to facilitate rural road surface disease target detection and automatic recognition. The self-developed automatic recognition system significantly enhances the accuracy and efficiency of pavement disease recognition. Furthermore, a management platform has been implemented to enable the dynamic management of rural road disease data and maintenance operations.

Keywords: rural road; distress sample annotation; automatic recognition; the U-Net algorithm; lightweight

1. Introduction

Rural roads are an important part of the road network, serving as a vital infrastructure for the economic development of rural areas. They are also critical to promoting urban–rural integration and driving rural revitalization. Because of the expansion of rural road coverage and accessibility, the management philosophy of rural roads has gradually shifted from focusing on construction to a coordinated development of construction, management, maintenance, and operation. However, the traditional rural road maintenance management model remains outdated and cannot meet the actual development needs of rural road operations.

With the rapid development of big data and AI technologies, various lightweight and automated devices have been applied to technically assess the conditions of rural roads [1]. Some regions have improved their management systems by constructing mathematical models to address road maintenance planning and traffic congestion issues [2,3]. Previous studies have optimized multi-objective identification, prioritizing safety hazard factors while meeting maintenance planning requirements [4]. Through these technological advancements, efforts are being made to promote the digitalization of infrastructure, specialization in maintenance, modernization of management, operational efficiency, and high-quality service, thereby driving the digital transformation of rural roads.

 https://doi.org/10.3390/buildings15234222

The condition assessment of rural roads and the evaluation of infrastructure service conditions are fundamental tasks in road construction, maintenance, and management [5]. By utilizing more lightweight and automated inspection methods, not only can road engineering quality be effectively controlled, but it can also be integrated with smart management systems to enhance the scientific level of road asset management and maintenance decision-making, further achieving the goal of "every road must be maintained, and maintenance must be in place."

In recent years, disease and defect detection methods have undergone significant technological advancements and innovations across various fields. In the area of concrete crack detection, Hu et al. [6] proposed a three-step algorithmic framework based on computer vision, which effectively identifies cracks on concrete surfaces even under complex background conditions. Hui et al. [7] applied the MobileNetV2 neural network for crack detection, significantly enhancing accuracy while enabling precise evaluation of crack widths in high-resolution images.

For bridge defect detection, Li et al. [8] developed a crack detection method incorporating geometric correction and calibration algorithms, addressing key challenges in UAV-based crack identification for bridge structures. An et al. [9] introduced an improved bridge deck defect detection algorithm based on YOLOv7-Tiny-DBB, which not only improved detection accuracy but also mitigated the issue of missed small-scale defects.

In terms of lightweight detection frameworks, Sheng et al. [10] proposed a maturity detection method for Hemerocallis citrina Baroni based on GCB YOLOv7, leveraging a lightweight neural network and attention mechanism to balance computational efficiency and detection precision. Jiang et al. [11] introduced MobiLiteNet, a lightweight deep learning model that significantly reduced computational overhead while delivering real-time, scalable, and accurate road hazard detection—supporting the advancement of intelligent transportation systems and infrastructure management.

Regarding road surface defect detection, France's GERPHO system [12] utilizes photographic technology combined with vehicle positioning to synchronously capture images of road damage, setting a new benchmark in the field. Similarly, Japan's Komatsu system [13] employs high-speed image acquisition with auxiliary lighting to collect high-resolution road surface damage data. The integration of digital imaging and computer vision technologies has given rise to rapid road surface inspection systems centered around digital cameras. The U.S. DHDV data collection vehicle [13] enables real-time, automated analysis through high-speed, multi-source data acquisition—marking a major step forward in automated road condition monitoring. These studies have provided successful performance for infrastructural disease and defect detection. However, an automatic identification system for rural road diseases and maintenance management is still rare. Despite the development and application of numerous pavement distress detection technologies, rural road environments pose significant challenges. Traditional detection systems are often large, expensive, and limited in both accuracy and processing speed, which hinders their effectiveness in supporting timely and cost-efficient rural road maintenance.

In this study, we aim to develop an AI-based, lightweight, low-cost, accurate, and efficient automatic identification system specifically designed for detecting rural road diseases and facilitating effective maintenance management. The primary objective is to integrate advanced machine learning and image processing techniques to improve the efficiency and accuracy of rural road condition assessments, which traditionally rely on manual inspections that are time-consuming and subjective. The study begins by detailing the overall design framework of the automatic identification system for rural road diseases. This includes an overview of the data acquisition process, system architecture, and the integration of AI modules with geographic information systems (GIS) and mobile platforms for real-time monitoring. Next, we introduce a robust rural road pavement distress automatic identification algorithm. This algorithm utilizes deep learning methods to analyze road surface images and classify various types of pavement distresses such as cracks, potholes, broken slabs, and so on. The methodology includes data preprocessing, model training, and performance evaluation, ensuring the system's reliability across different rural environments and lighting conditions. Section 4 presents practical applications of the developed system in rural road maintenance and management. We demonstrate how the system supports decision-making by local authorities through automated reporting, prioritization of repairs, and maintenance scheduling based on severity levels and traffic patterns. Finally, Section 5 summarizes the key findings and contributions of the study. It highlights the effectiveness of the AI-based approach in reducing maintenance costs, improving road safety, and promoting sustainable infrastructure development in rural areas.

2. Design of the Automatic Identification System for Rural Road Diseases

2.1. System Architecture Design

The system architecture mainly consists of five layers: the Acquisition Layer, Network Layer, Data Layer, Application Layer, and Platform Layer, as shown in Figure 1.

The Acquisition Layer aims to achieve full coverage of automatic inspections for rural roads. It uses lightweight automated detection devices to real-time collect basic location information of the current vehicle (such as route code, route name, pile number, maintenance unit, etc.), as well as pavement and forward image data, ensuring efficient data collection at the grassroots level of rural roads.

The Network Layer takes into account the communication network conditions of rural roads. It uses industrial computers to self-check network status and automatically switches between transmission and storage modes based on current signal strength and network bandwidth, ensuring the integrity and continuity of underlying data.

The Data Layer utilizes artificial intelligence algorithms to analyze and classify the underlying data, establishing a multidimensional database that includes the basic information database, distress information database, road environment database, inspection information database, geographic information database, and maintenance information database.

The Application Layer is the core of the intelligent management and maintenance system for rural roads. It integrates and analyzes multidimensional data to build corresponding functionalities that meet the practical needs of rural road maintenance and management. It mainly includes data analysis, result evaluation, and decision support.

The Platform Layer logically connects various data modules within the system and displays and operates them on the Web and mobile apps in the form of image models, reports, and other formats.

2.2. Design of Lightweight Intelligent Inspection Hardware System

The lightweight intelligent inspection hardware system is an integrated hardware solution comprising a high-definition camera, a multi-mode geolocation positioning module, a central control panel, a 4G/5G network module, and an edge intelligence all-in-one unit. The system is magnetically mounted for easy installation and removal.

Each set of the lightweight intelligent inspection hardware system includes one high-definition industrial camera; one edge intelligence all-in-one unit (consisting of one industrial computer, one 4G/5G transmission module, one GPS positioning module, and four antennas); one data acquisition and control display unit; and several power connection and data cables. The system configuration and device technical specifications are shown in Figure 2 and Table 1.

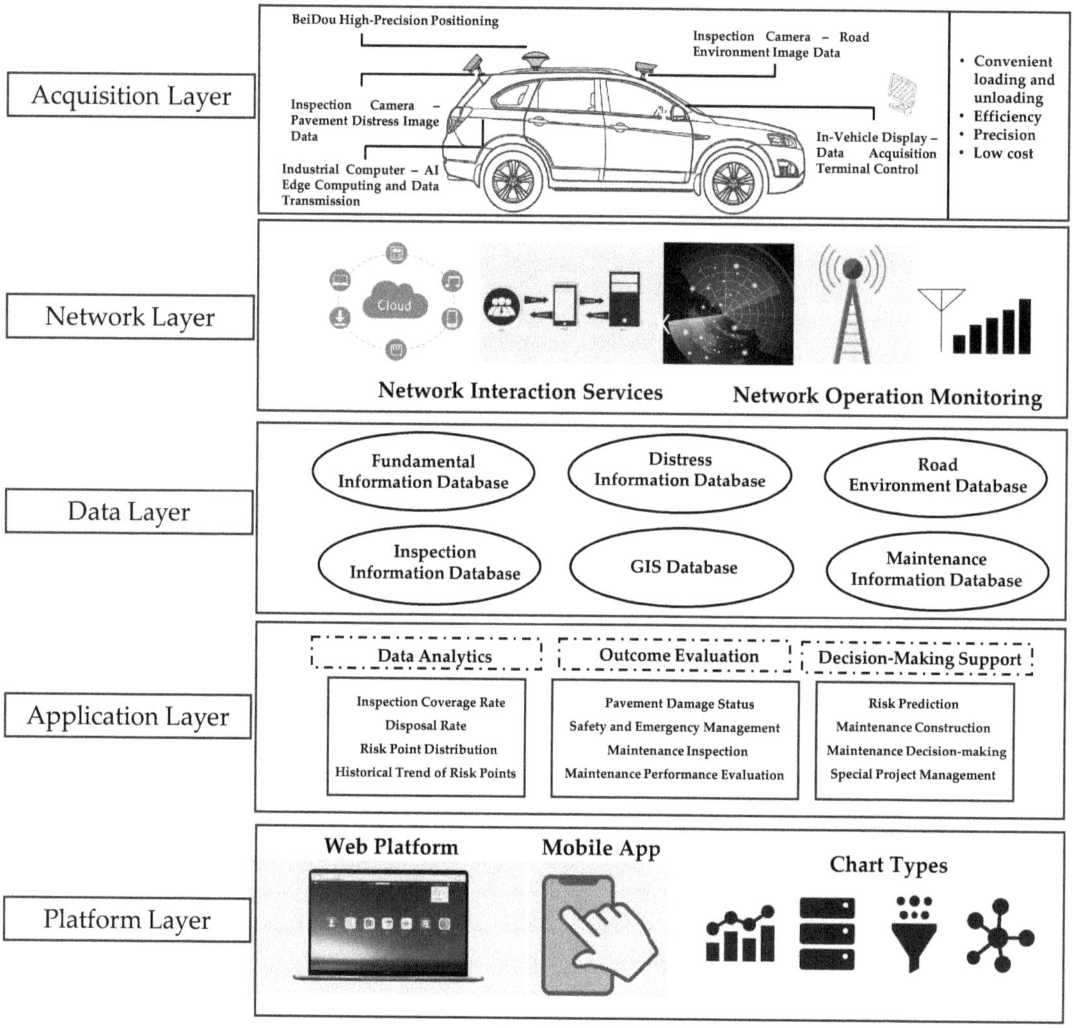

Figure 1. System architecture design.

Table 1. Technical Specifications.

Item	Description
CPU	11th Gen Inter(R) Core(TM) I5-1135G7@2.49GHZ
RAM/ROM	8 GB-DDR3L 1600 MHz
Operation System	Windows 11 Professional Edition 22H2
Hard Drive	Samsung SSD 870 EVO 500GB
Ethernet Port	4× Intel® i211 Gigabit Ethernet Ports; supports PoE (Power over Ethernet); compliant with IEEE 802.3af; supports Wake-on-LAN/PXE boot

Table 1. *Cont.*

Item	Description
COM Port	4× RS232/RS485, switchable via jumper caps; COM1–COM4 support digital capacitive isolation
Wireless Communication	1× External SIM card slot connected to M.2 B-key slot, supports 4G/5G wireless networking
GPS	GPS + BDS + SBAS + QZSS hybrid positioning with <2 m positioning error and <0.1 m/s velocity error; output frequency: 10 Hz
Camera	Triangular magnetic base; 2560 × 1440 resolution; autofocus; strong light suppression; backlight enhancement; Frame rate: 50 Hz, 25 fps
Power Connection	Connects to vehicle power supply
Power Supply Voltage	24 V, 5 A
Operating Temperature	−20 to +70 °C
Waterproof Rating	IP67-rated waterproof
Antenna Type	4× High-gain magnetic metal rod antennas
Antenna Height	230 mm ± 3 mm
Device Installation Method	The GPS module and camera are magnetically mounted; the industrial computer is fixed in an enclosed housing

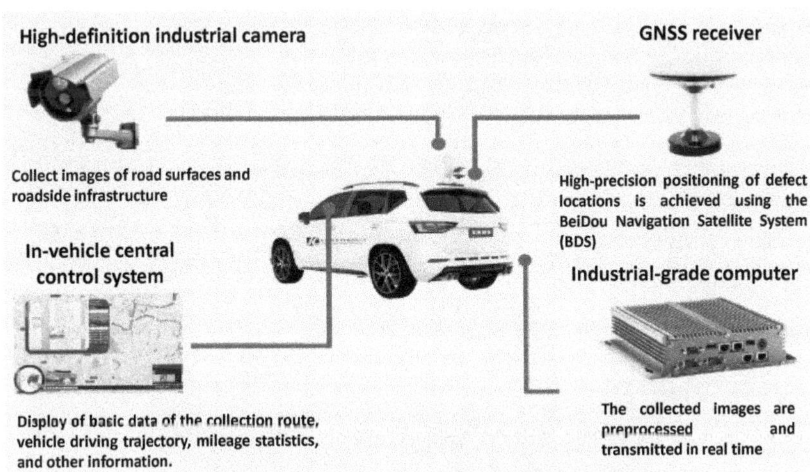

Figure 2. Composition of the lightweight intelligent inspection hardware system.

3. Automatic Identification Algorithm for Rural Road Pavement Distress

3.1. Typical Types of Rural Road Pavement Distress

The classification of pavement distresses on rural roads differs from that of standard trunk roads. The detection environment of regular trunk roads is favorable, and the identification of distress types is relatively accurate, whereas the detection environment of rural roads is complex, making it difficult to distinguish between different types of distress. Common pavement distress types frequently observed on rural roads include transverse cracks, longitudinal cracks, alligator cracking, and potholes for asphalt pavements, while cement concrete pavements typically exhibit cracks, broken slabs, and potholes. For the purpose of automatic identification of pavement distresses on rural roads, accurate segmentation of the damaged areas is essential. By statistically analyzing the extent of pavement damage, relevant evaluation indices can be derived. To enhance the precision in quantifying the affected areas, this study categorizes typical rural pavement distresses into two groups: length-based and area-based. Transverse cracks, longitudinal cracks, and general cracks are classified as length-based distresses, whereas alligator cracking, broken slabs, and potholes fall under area-based distresses.

3.2. Distress Sample Annotation Method

High-quality road distress samples serve as the foundation for developing effective algorithmic recognition models, and the choice of annotation method significantly influences the subsequent training and performance of deep learning models. At present, pavement distress annotation primarily relies on manual labeling. In addition to mastering the operation of annotation software, annotators must possess relevant expertise in road inspection to accurately identify various types of pavement diseases.

Currently, manual annotation methods mainly include the bounding box method, grid method, and contour method, as illustrated in Figures 3 and 4. The bounding box method involves drawing an external rectangular box tangential to the edges of the pavement distress to estimate the affected area. The grid method divides pavement image data using 0.1×0.1 m grid lines, determines whether each grid cell falls within the diseased area, and calculates the number of affected grid cells to measure the distress area. The contour method involves tracing the actual edges of the pavement distress to form a closed shape, and the area enclosed by the traced contour is used to quantify the distress extent.

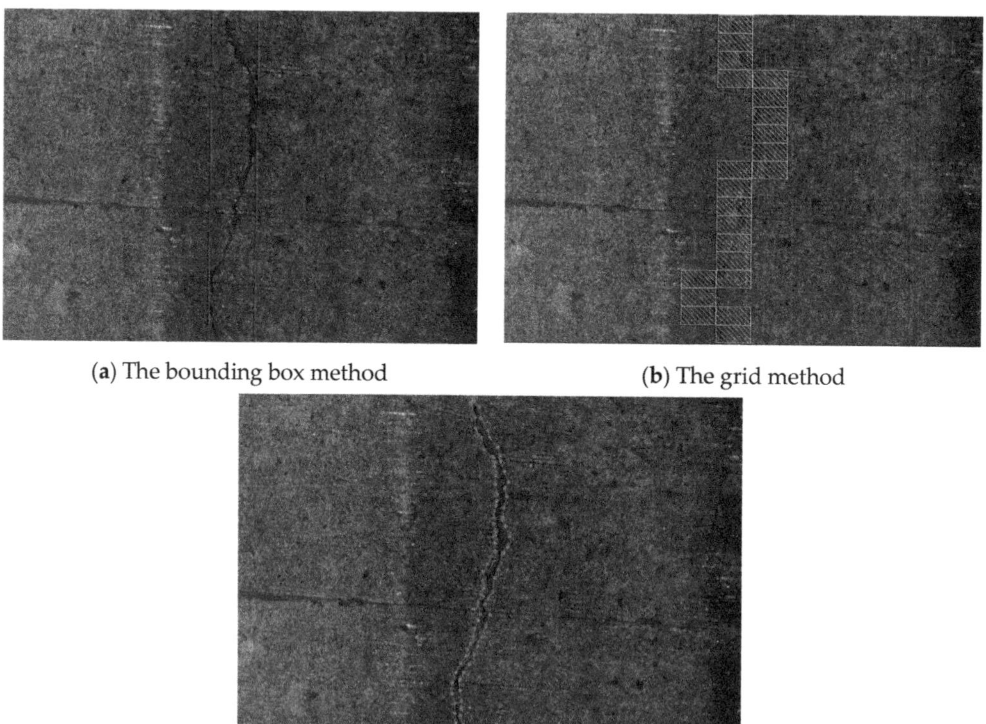

(**a**) The bounding box method (**b**) The grid method

(**c**) The contour method

Figure 3. Annotated examples of length-based distress samples.

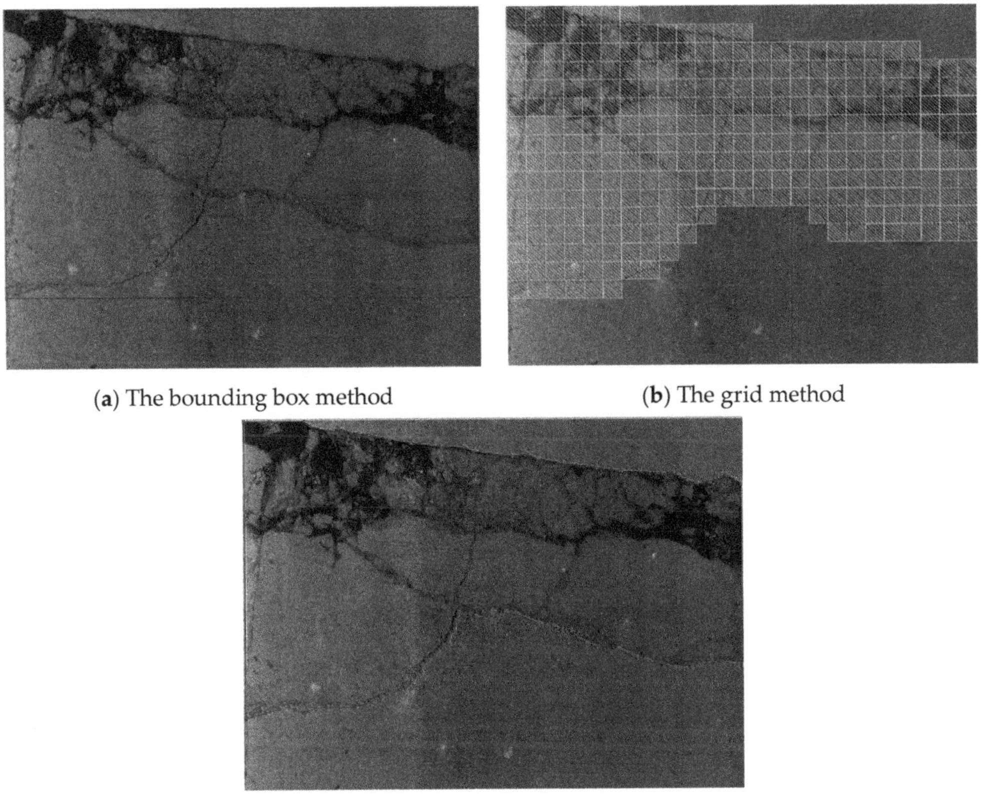

(a) The bounding box method (b) The grid method

(c) The contour method

Figure 4. Annotated examples of area-based distress samples.

3.3. *Typical Detection and Identification Process of Rural Road Diseases*

The automatic detection and identification system for road diseases in this paper is based on a disease segmentation algorithm, conducting research on the identification of typical diseases in rural roads. The main recognition process is divided into the following three steps:

Step 1: Random samples are taken from rural road pavement image data in different regions, with varying technical levels and pavement types in Zhejiang Province. The images are annotated for disease identification to build a typical rural road disease sample dataset.

Step 2: Optimizations are made based on the U-Net to construct the disease type localization and accurate segmentation of disease areas.

Step 3: The deep learning algorithm model is tested to evaluate the recognition performance of typical rural road pavement diseases in the test database.

3.3.1. Establishing a Sample Database of Typical Rural Road Distress

Image data of rural road surfaces in various regions of Zhejiang Province, including Hangzhou, Ningbo, Shaoxing, Quzhou, Taizhou, Lishui, and Jinhua, were collected using lightweight equipment. Distressed images were randomly selected from roads with different technical grades and pavement types to construct a typical rural road distress sample dataset. The asphalt pavement dataset primarily includes longitudinal cracks, transverse cracks, alligator cracking, and potholes, while the cement pavement dataset includes cracks, shattered slabs, and potholes. These distress types were used as the basis for annotation. To ensure the quality of the distress sample data annotation, preprocessing steps were carried out on the raw image data, which included noise reduction, resizing, and adjustments to brightness and contrast.

In this study, the U-Net, an advanced architecture derived from the Fully Convolutional Network (FCN), is employed. It features a symmetric encoder–decoder structure with extensive skip connections, which effectively preserve spatial information and improve segmentation accuracy. The U-Net disease detection algorithm presented in this study classifies asphalt pavement cracks into transverse and longitudinal cracks. The specific classification criterion involves calculating the crack angle based on the long edge of the minimum bounding rectangle and defining the angle range: cracks with angles between 0° and 45° are classified as transverse cracks, while those with angles between 45° and 90° are classified as longitudinal cracks.

As shown in Figure 5a, the lengths of the long and short edges of the bounding rectangle are first determined (with d1 representing the long edge and d2 representing the short edge). Then, the angle between the long edge and the horizontal axis (x) is calculated. Two angles, $\alpha 1$ and $\alpha 2$, are obtained, and the smaller angle, which is $\leq 90°$, is selected (as shown in the figure, if $\alpha 1$ is less than 90°, this value is chosen). Figure 5b provides an example from a rural road image. The minimum bounding rectangle of the measured pavement crack has an angle of 27°, measured using a protractor, indicating it is a transverse crack.

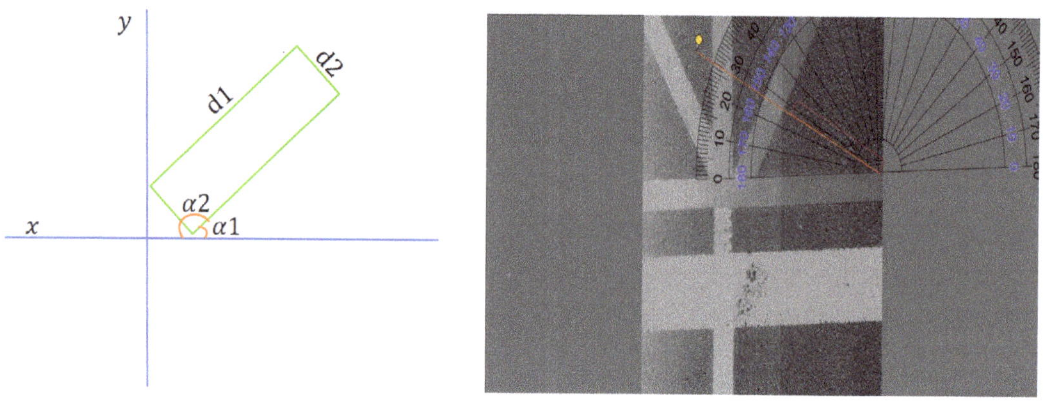

(**a**) Schematic diagram of a bounding rectangle

(**b**) Example of transverse cracks

Figure 5. Example of the bounding rectangle to determine the orientation of the crack.

The annotation method for the typical rural road distress samples is the "contour method," which requires that sample image data be clear and unobstructed. Images that are blurred or difficult to interpret for crack type classification should be excluded. The rural road typical distress sample database includes 1487 images of asphalt transverse cracks, 3654 images of asphalt longitudinal cracks, 2951 images of asphalt alligator cracks, 817 images of asphalt potholes, 6755 images of cement cracks, 4875 images of cement shattered slabs, and 1812 images of cement potholes, as shown in Figure 6. Table 2 presents the feature list of the typical rural road distress samples, which serves as the standard for annotation and provides a foundation for the development of automated distress recognition algorithms.

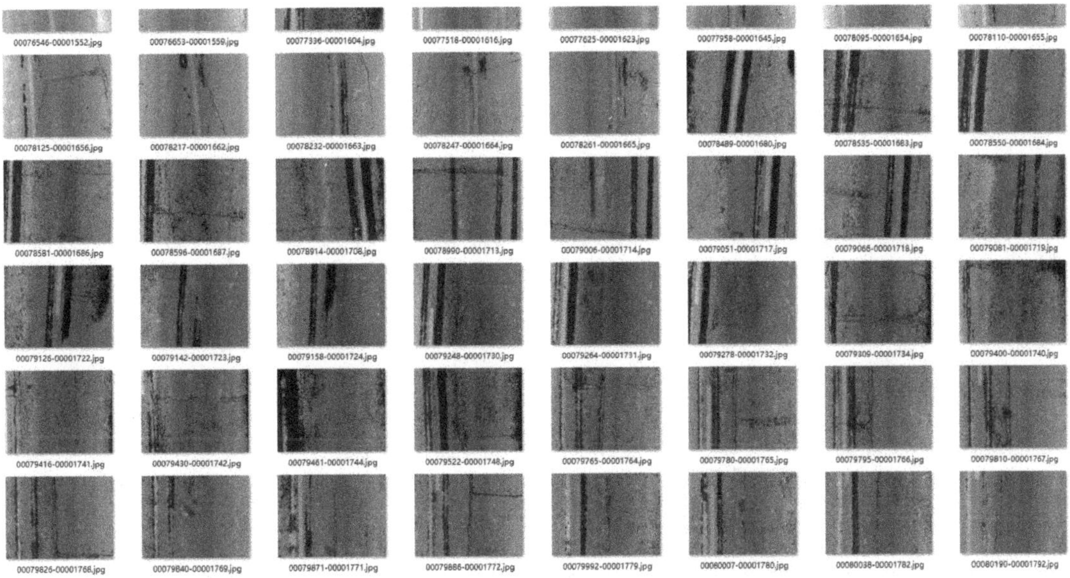

Figure 6. Sample images from the database.

Table 2. Feature table of typical rural road distress.

Annotated Images	Type	Feature
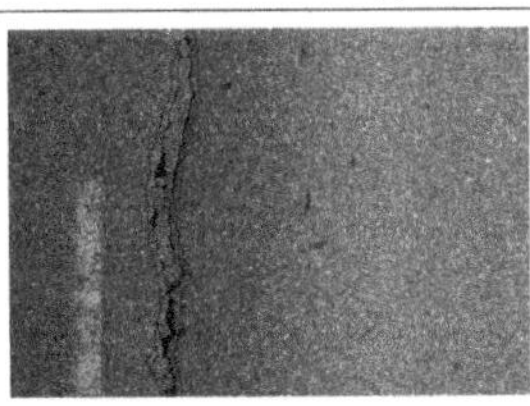	Longitudinal cracks in asphalt pavements	The cracks are generally parallel to the driving direction, with the angle $\alpha1$ clearly falling between 45° and 90°. The annotation even includes fine cracks, with markings along the outer edge of the crack.
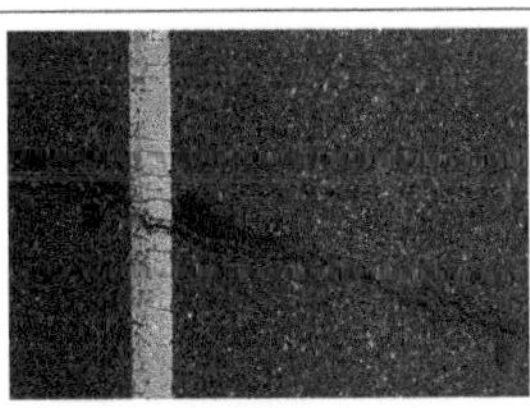	Transverse cracks in asphalt pavements	The cracks are generally perpendicular to the driving direction, with the angle $\alpha1$ clearly falling between 0° and 45°. If the crack is within a strip repair area, the boundary between the repair and the crack should be identified. Markings are made along the outer edge of the crack.
	Alligator cracking in asphalt pavements	Alligator cracking on asphalt pavement shall be defined as a grid-shaped pattern formed by intersecting longitudinal and transverse cracks. Annotations must capture the fine internal cracking within the network, with delineation traced along the outermost boundary of the cracking zone.

Table 2. *Cont.*

Annotated Images	Type	Feature
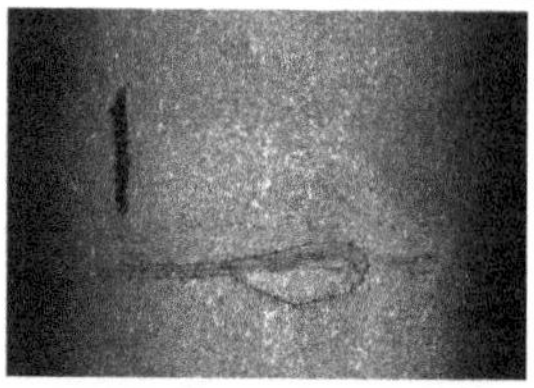	Pothole in asphalt pavements	Potholes are localized surface depressions resulting from the disintegration and loss of aggregate in asphalt pavement. Markings should be made along the outer edge of the pothole.
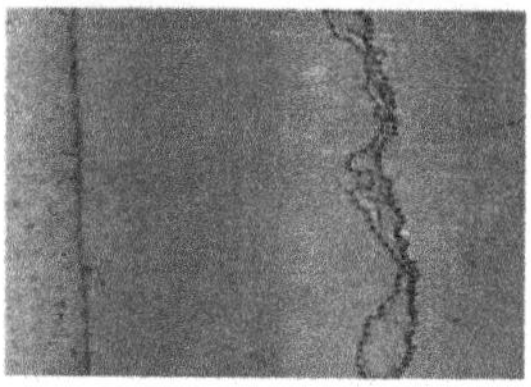	Cracks in concrete pavements	The cracks are longitudinal, transverse, or diagonal cracks on the cement pavements. The annotation should include fine cracks, marked along the outer edge of the crack.
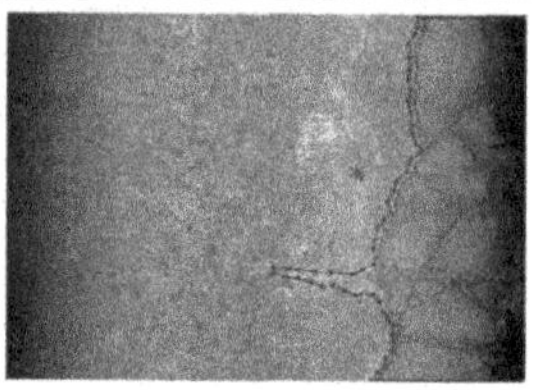	Broken Slab in concrete pavements	The fractured slab refers to the cement pavement slab with through cracks penetrating the surface layer, and the slab is divided into three or more pieces by the cracks. The annotation should include fine cracks within the range of the through cracks, marked along the outer edge of the fractured slab.
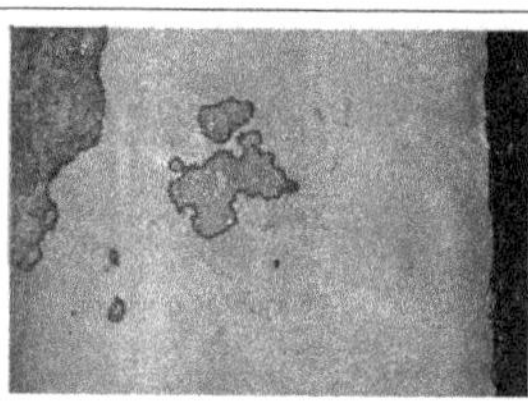	Pothole in concrete pavements	Potholes refer to localized depressions or damage on the cement pavement. The annotation should determine whether the potholes are independent. Independent potholes should be marked along their outer edges separately.
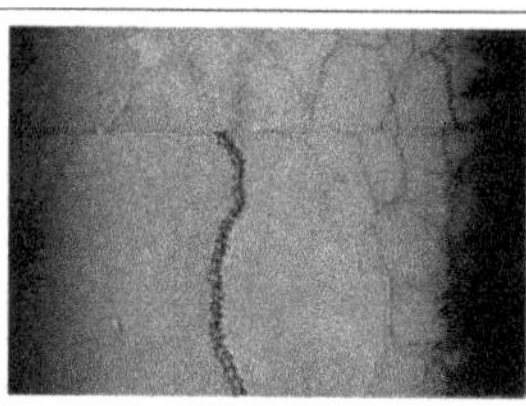	Combined pavement distresses 1	If independent cracks outside the fractured slab area are observed in the pavement image, they should be annotated according to the type of defect. When other defects appear within the fractured slab, only the fractured slab should be marked.
	Combined pavement distresses 2	If independent potholes outside the extent of the alligator cracking are observed in the pavement image, they should be annotated according to the type of defect. When other defects appear within the alligator cracking area, only the alligator cracking should be marked.

3.3.2. Construction of a Typical Rural Road Disease Identification Algorithm

In this study, the U-Net, an advanced architecture derived from the Fully Convolutional Network (FCN), is employed. It features a symmetric encoder–decoder structure with extensive skip connections, which effectively preserve spatial information and improve segmentation accuracy. The U-Net consists of two concatenated paths: the contracting path and the expanding path. The contracting path is designed to extract image features, compressing the image into a feature map that represents these features. The expanding path is responsible for precise localization, decoding the extracted features into a segmented prediction image that matches the original image size.

Unlike the FCN, U-Net retains a large number of feature channels during the upsampling process, allowing more information to be retained and flow into the final reconstructed segmentation image. To minimize information loss in the contracting path, feature maps of the same size from both the contracting and expanding paths are concatenated, followed by convolution and upsampling. This process integrates more information, thereby enhancing the accuracy of image segmentation. The U-Net network structure is illustrated in Figure 7.

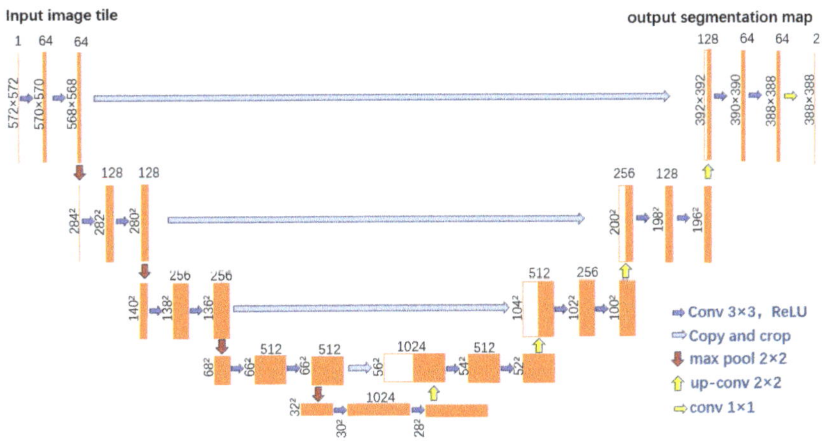

Figure 7. U-Net network structure.

3.3.3. Recognition Results Under Different Annotation Methods

A deep learning-based U-Net algorithm model was used to evaluate the recognition performance of typical rural road distresses through metrics such as accuracy, false detection rate, and missed detection rate. Using lightweight inspection equipment, rural roads within the Hangzhou area were surveyed, resulting in the selection and classification of 28,000 test samples across 7 typical types of rural road distress. Experimental parameters are provided in Table 3. The recognition results are shown in Table 4, and examples of recognition images are presented in Table 5.

Table 3. Experimental parameters.

Experimental Parameters	Content
Dataset Proportion	Training:Validation = 8:2
Model Parameters	Batch Size: 8, Learning Rate: 0.0001, Optimizer: SGD, Number of Epochs: 100, Loss Function: CrossEntropyLoss

The statistical results indicate that the U-Net algorithm trained using the contour method outperforms both the grid method and the bounding box method in terms of accuracy, false detection rate, and missed detection rate. The contour method achieves the best performance, with an average accuracy of 94.95%, an average false detection rate of 5.93%, and an average missed detection rate of 2.79%, as shown in Table 4.

Table 4. Recognition results for different annotation methods.

Annotation Method	Road Type	Distress Type	Sample Images	Accuracy	False Detection	Missed Detection	Accuracy (%)	False Detection Rate (%)	Missed Detection Rate (%)
Bounding box method	Asphalt pavement	Longitudinal cracks	4000	3792	123	85	94.79%	4.07%	2.87%
		Transverse cracks	4000	3780	122	99	94.49%	5.12%	4.20%
		Alligator cracking	3000	2847	138	15	94.89%	9.52%	1.16%
		Potholes	3000	2710	233	58	90.32%	10.65%	2.87%
	Concrete pavements	Cracks	8000	7694	223	83	96.18%	4.26%	1.64%
		Broken Slab	3000	2827	125	48	94.23%	10.05%	4.07%
		Potholes	3000	2836	51	112	94.55%	3.39%	7.14%
Average							94.20%	6.72%	3.42%
Grid method	Asphalt pavement	Longitudinal cracks	4000	3809	113	78	95.22%	3.75%	2.61%
		Transverse cracks	4000	3790	116	94	94.75%	4.89%	3.97%
		Alligator cracking	3000	2859	130	12	95.29%	8.93%	0.87%
		Potholes	3000	2726	224	50	90.86%	10.23%	2.50%
	Concrete pavements	Cracks	8000	7704	216	80	96.30%	4.14%	1.57%
		Broken Slab	3000	2841	118	41	94.69%	9.45%	3.53%
		Potholes	3000	2851	43	106	95.04%	2.84%	6.72%
Average							94.59%	6.32%	3.11%
Contour method	Asphalt pavement	Longitudinal cracks	4000	3821	107	72	95.53%	3.53%	2.40%
		Transverse cracks	4000	3804	110	86	95.10%	4.60%	3.63%
		Alligator cracking	3000	2869	124	7	95.63%	8.50%	0.52%
		Potholes	3000	2741	216	43	91.37%	9.85%	2.13%
	Concrete pavements	Cracks	8000	7713	211	76	96.41%	4.03%	1.49%
		Broken Slab	3000	2853	109	38	95.10%	8.67%	3.20%
		Potholes	3000	2866	36	98	95.53%	2.36%	6.18%
Average							94.95%	5.93%	2.79%

Table 5. Image of typical defect identification in rural roads.

Road Type	Distress Type	Defect Identification Example Image 1	Defect Identification Example Image 2
Asphalt pavements	Longitudinal cracks	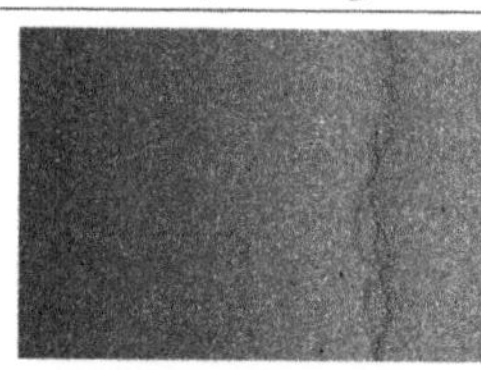	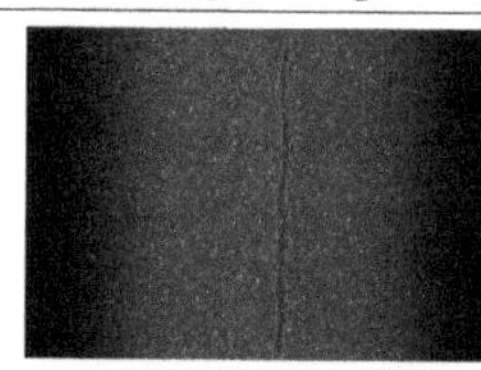
	Transverse cracks		
	Alligator cracking	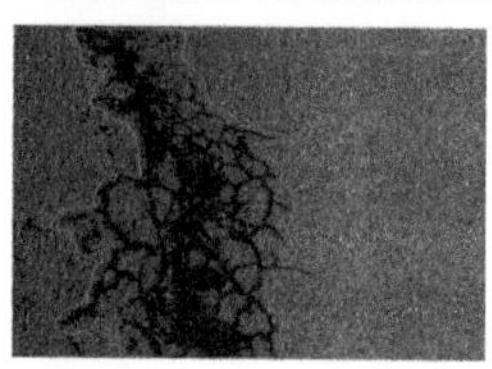	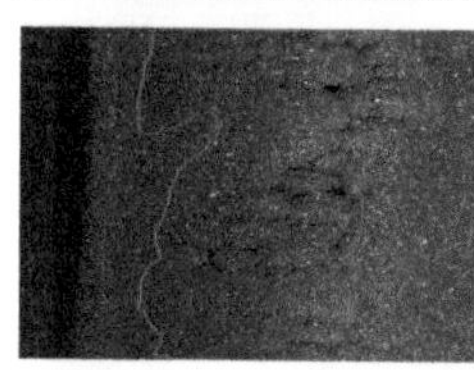
	Potholes	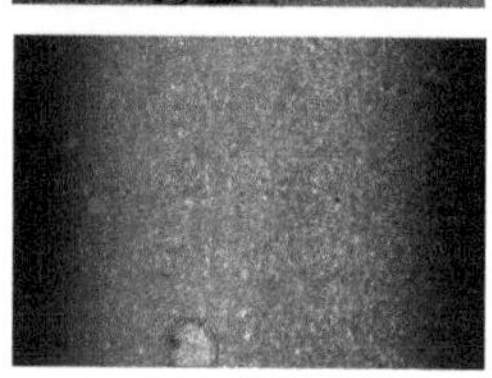	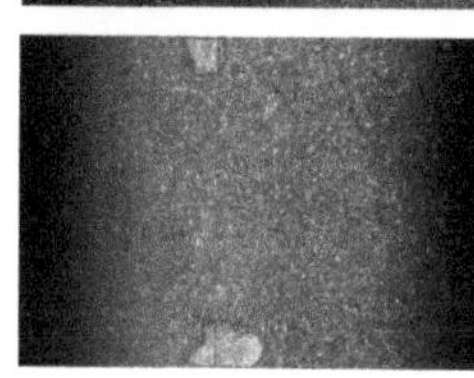

Table 5. *Cont.*

Road Type	Distress Type	Defect Identification Example Image 1	Defect Identification Example Image 2
Concrete pavements	Cracks		
	Broken Slab	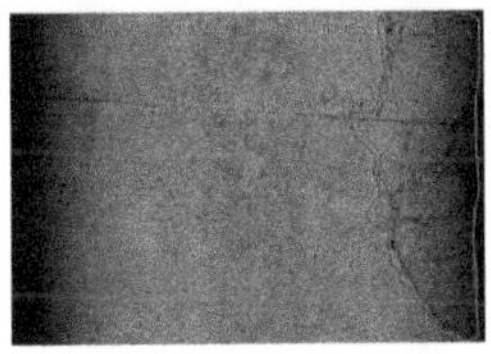	
	Potholes	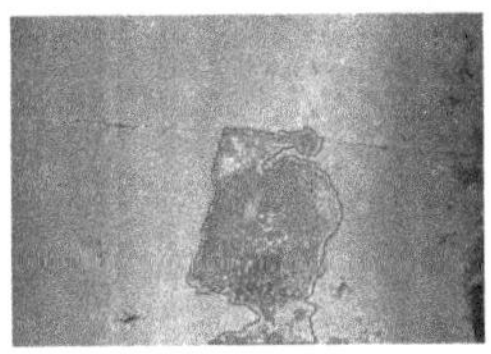	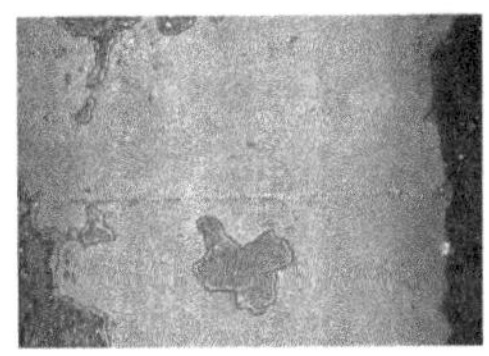

3.3.4. Comparison of Automatic Distress Recognition Models for Different Pavement Types

To further validate the accuracy and efficiency of the proposed model, a comparative experiment using different automatic recognition models was conducted to evaluate algorithm performance. The most representative models selected for comparison were the YOLO v3 and Faster R-CNN automatic recognition models.

The test data consisted of approximately 10 km of rural road surface images collected using lightweight inspection equipment in Tonglu District, Hangzhou. For seven types of typical pavement distresses, three recognition models—U-Net-based model, YOLO v3 [14], and Faster R-CNN [15]—were applied to the same dataset for performance comparison, as shown in Table 6.

Table 6. Comparison of evaluation metrics for different recognition models.

Model	Evaluation Metrics		
	Accuracy Rate	False Detection Rate	Missed Detection Rate
YOLO v3	90.36%	6.74%	2.9%
Fastr R-CNN	93.86%	4.74%	1.4%
Unet	95.04%	3.67%	1.29%

The statistical results indicate that the proposed U-Net-based model outperformed the other models in terms of accuracy, false detection rate, and missed detection rate across all distress types.

4. Application for Rural Road Maintenance and Management

Based on an automatic identification algorithm developed for rural road distresses, a rural road disease management system that enables dynamic management of rural road maintenance operations is developed. Its key functions related to the above automatic identification algorithm are outlined below:

(1) Data validation

The inspection management platform enables the review of completed image recognition data, allowing the identification and filtering of inaccurate results (see Figure 8). Through the system's editing functions, technical personnel can modify or delete distress annotations to ensure the accuracy and reliability of the recognition outcomes.

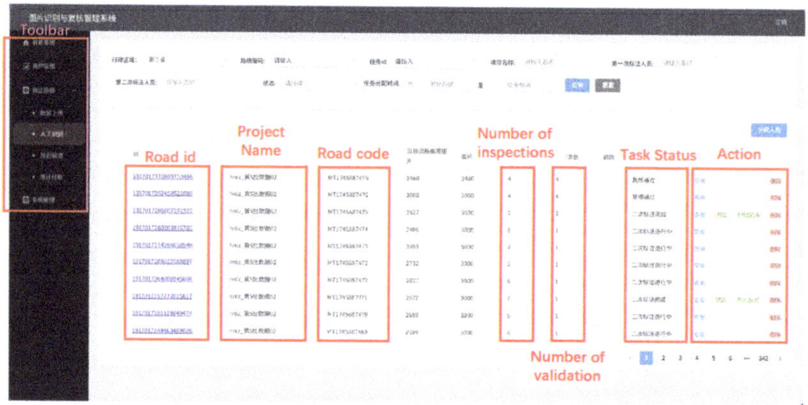

Figure 8. Data validation module.

(2) Rural road distress database

Through the inspection management platform, verified recognition results are categorized and stored, laying a solid foundation for future iterations and upgrades of the distress recognition algorithm, while also supporting historical distress query requirements for inspection projects.

(3) Maintenance Work Orders

The inspection management platform enables rapid identification of rural road surface distress. Based on the identified distress types, maintenance work orders, as shown in Figure 9, are generated in real time, significantly reducing the time span between distress detection and maintenance dispatch, and greatly improving the efficiency of rural road maintenance.

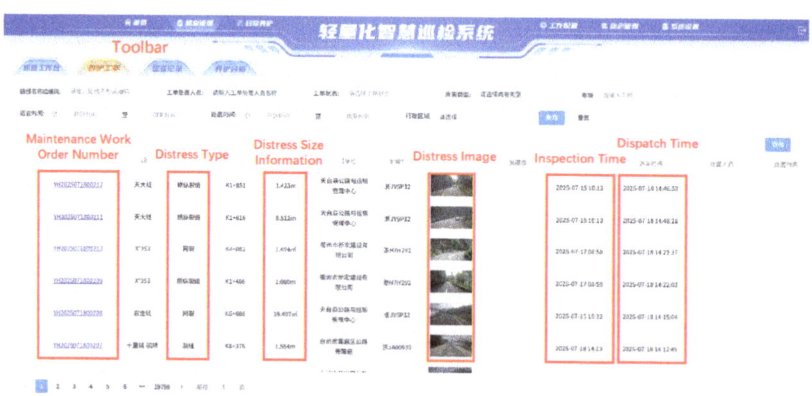

Figure 9. Maintenance work orders.

(4) Data Reports

According to the recognition results from the inspection management platform, data reports in Excel Format, as shown in Figure 10, on rural road surface distresses are automatically generated. The reports provide detailed information, including distress locations, affected areas, and distress types, serving as a foundation for subsequent maintenance analysis.

路线编码	路线名称	展示桩号	车牌号	病害类型名称	病害尺寸信息	巡查时间	行政区划	设备业主
X311330681	石东线	K10+364	浙D5M6N6	裂缝	4.410m	7/7/25 9:43 AM	诸暨市	绍兴市诸暨市璜山公路养护站
S211	桐洞线	K13+920	浙FZZ709	横纵裂缝	3.274m	7/7/25 9:42 AM	桐乡市	嘉兴市公路与运输管理中心
X317330681	三联线	K0+969	浙D7S3N7	裂缝	15.652m	7/7/25 9:41 AM	诸暨市	绍兴市诸暨市江藻公路养护站
X317330681	三联线	K1+418	浙D7S3N7	网裂	11.353m²	7/7/25 9:41 AM	诸暨市	绍兴市诸暨市江藻公路养护站
S211	桐洞线	K13+365	浙FZZ709	横纵裂缝	2.013m	7/7/25 9:40 AM	桐乡市	嘉兴市公路与运输管理中心
S211	桐洞线	K12+465	浙FZZ709	网裂	6.267m²	7/7/25 9:40 AM	桐乡市	嘉兴市公路与运输管理中心
S211	桐洞线	K12+550	浙FZZ709	网裂	10.334m²	7/7/25 9:40 AM	桐乡市	嘉兴市公路与运输管理中心
S211	桐洞线	K12+565	浙FZZ709	横纵裂缝	15.262m	7/7/25 9:40 AM	桐乡市	嘉兴市公路与运输管理中心
S211	桐洞线	K12+610	浙FZZ709	横纵裂缝	4.531m	7/7/25 9:39 AM	桐乡市	嘉兴市公路与运输管理中心
S211	桐洞线	K13+025	浙FZZ709	横纵裂缝	2.302m	7/7/25 9:39 AM	桐乡市	嘉兴市公路与运输管理中心
S211	桐洞线	K12+650	浙FZZ709	横纵裂缝	8.441m	7/7/25 9:39 AM	桐乡市	嘉兴市公路与运输管理中心
S211	桐洞线	K12+660	浙FZZ709	横纵裂缝	19.727m	7/7/25 9:39 AM	桐乡市	嘉兴市公路与运输管理中心
S211	桐洞线	K12+660	浙FZZ709	横纵裂缝	4.584m	7/7/25 9:39 AM	桐乡市	嘉兴市公路与运输管理中心
S211	桐洞线	K12+675	浙FZZ709	横纵裂缝	11.521m	7/7/25 9:39 AM	桐乡市	嘉兴市公路与运输管理中心
S211	桐洞线	K12+680	浙FZZ709	横纵裂缝	5.168m	7/7/25 9:39 AM	桐乡市	嘉兴市公路与运输管理中心
S211	桐洞线	K12+875	浙FZZ709	网裂	7.430m²	7/7/25 9:39 AM	桐乡市	嘉兴市公路与运输管理中心
S211	桐洞线	K12+860	浙FZZ709	横纵裂缝	5.857m	7/7/25 9:39 AM	桐乡市	嘉兴市公路与运输管理中心
S211	桐洞线	K12+830	浙FZZ709	横纵裂缝	2.075m	7/7/25 9:39 AM	桐乡市	嘉兴市公路与运输管理中心
S211	桐洞线	K12+690	浙FZZ709	横纵裂缝	6.644m	7/7/25 9:39 AM	桐乡市	嘉兴市公路与运输管理中心
S211	桐洞线	K12+705	浙FZZ709	横纵裂缝	4.559m	7/7/25 9:39 AM	桐乡市	嘉兴市公路与运输管理中心
S211	桐洞线	K12+715	浙FZZ709	横纵裂缝	4.108m	7/7/25 9:39 AM	桐乡市	嘉兴市公路与运输管理中心
S211	桐洞线	K12+665	浙FZZ709	横纵裂缝	5.369m	7/7/25 9:39 AM	桐乡市	嘉兴市公路与运输管理中心
S211	桐洞线	K12+645	浙FZZ709	横纵裂缝	18.824m	7/7/25 9:39 AM	桐乡市	嘉兴市公路与运输管理中心
S211	桐洞线	K12+640	浙FZZ709	横纵裂缝	16.742m	7/7/25 9:39 AM	桐乡市	嘉兴市公路与运输管理中心

Figure 10. Data report in Excel format (in Chinese).

5. Conclusions

This study addresses the detection needs of rural road diseases by designing an automatic recognition system based on a disease segmentation algorithm. The main novelty lies in the creation of a typical rural road disease sample database covering different regions, technical grades, and pavement types in Zhejiang Province. Additionally, a pavement disease recognition network is optimized and built based on the U-Net algorithm. By comparing the model training results of three annotation strategies, the optimal annotation method is determined. The results show that the U-Net algorithm trained using the contour annotation method achieves the best performance, with an accuracy of 94.95%, a false detection rate of 5.93%, and a missed detection rate of 2.79%, significantly outperforming the grid and bounding box methods. The system has completed initial deployment and debugging, enabling efficient data processing for lightweight detection devices. Under controlled testing conditions, the system demonstrates excellent performance, effectively addressing the challenges faced by traditional equipment in rural road detection. It provides technical support for low-cost and precise maintenance, showcasing its vast potential for future practical applications.

Author Contributions: Conceptualization, L.C. and H.Z.; methodology, H.Z. and D.L.; software, D.L. and Y.L.; validation, Y.L. and J.L.; formal analysis, J.L. and K.F.; data collection, K.F. and J.L.; writing—original draft preparation, L.C.; writing—review and editing, H.Z.; project administration, L.C.; funding acquisition, D.L. and H.Z. All authors have read and agreed to the published version of the manuscript.

Funding: This work was supported by Zhejiang Provincial Department of Transportation Science and Technology Program Project (No. 2024018).

Data Availability Statement: Dataset available on request from the authors.

Conflicts of Interest: The authors declare no conflicts of interest.

References

1. Zhu, J.; Wu, Y.; Ma, T. Multi-Object Detection for Daily Road Maintenance Inspection with UAV Based on Improved YOLOv8. *IEEE Trans. Intell. Transp. Syst.* **2024**, *25*, 16548–16560. [CrossRef]
2. Mathew, B.S.; Isaac, K.P. Optimisation of Maintenance Strategy for Rural Road Network Using Genetic Algorithm. *Int. J. Pavement Eng.* **2014**, *15*, 352–360. [CrossRef]
3. Shan, F.; Li, H.; Wang, Z.; Jin, M.; Chen, D. Optimizing Rural Highway Maintenance Scheme with Mathematical Programming. *Appl. Sci.* **2024**, *14*, 8253. [CrossRef]
4. Pasindu, H.; Ranawaka, R.; Sandamal, R.; Dias, T. Incorporating Road Safety into Rural Road Network Pavement Management. *Int. J. Pavement Eng.* **2021**, *23*, 4306–4319. [CrossRef]
5. Yang, G.; Wang, H.; Pan, Y.; Li, L. Research on Application of the Rural Road Performance Assessment Method in Jiangsu Province. *Road Mater. Pavement Des.* **2017**, *18* (Suppl. 3), 76–87. [CrossRef]
6. Hu, Y.; Chen, N.; Hou, Y.; Lin, X.; Jing, B.; Liu, P. Lightweight Deep Learning for Real-Time Road Distress Detection on Mobile Devices. *Nat. Commun.* **2025**, *16*, 4212. [CrossRef] [PubMed]
7. Hui, L.; Ibrahim, A.; Hindi, R. Computer Vision-Based Concrete Crack Identification Using MobileNetV2 Neural Network and Adaptive Thresholding. *Infrastructures* **2025**, *10*, 42. [CrossRef]
8. Li, J.; Li, X.; Liu, K.; Yao, Z. Crack Identification for Bridge Structures Using an Unmanned Aerial Vehicle (UAV) Incorporating Image Geometric Correction. *Buildings* **2022**, *12*, 1869. [CrossRef]
9. An, H.; Fan, Y.; Jiao, Z.; Liu, M. Research on Improved Bridge Surface Disease Detection Algorithm Based on YOLOv7-Tiny-DBB. *Appl. Sci.* **2025**, *15*, 3626. [CrossRef]
10. Sheng, B.; Wu, L.; Zhang, N. A Maturity Detection Method for Hemerocallis Citrina Baroni Based on Lightweight and Attention Mechanism. *Appl. Sci.* **2023**, *13*, 12043. [CrossRef]
11. Jiang, M.; Gielen, G.; Deng, B.; Zhu, X. A Fast Learning Algorithm for Time-Delay Neural Networks. *Inf. Sci.* **2002**, *148*, 27–39. [CrossRef]
12. Kim, J.Y. Development of New Automated Crack Measurement Algorithm Using Laser Images of Pavement Surface. Master's Thesis, The University of Iowa, Iowa City, IA, USA, 2008.
13. Wang, K.C.P.; Gong, W.-G. Real-Time Automated Survey System of Pavement Cracking in Parallel Environment. *J. Infrastruct. Syst.* **2005**, *11*, 154–164. [CrossRef]
14. Redmon, J.; Farhadi, A. YOLOv3: An Incremental Improvement. *arXiv* **2018**. [CrossRef]
15. Ren, S.; He, K.; Girshick, R.; Sun, J. Faster R-CNN: Towards Real-Time Object Detection with Region Proposal Networks. *IEEE Trans. Pattern Anal. Mach. Intell.* **2017**, *39*, 1137–1149. [CrossRef] [PubMed]

buildings

MDPI

Article

Stakeholders' Perception and Adoption of Upcycling for Material Utilisation Plans in Road Construction: The Case of South Africa

Salome Naicker [1,2], Mohamed Mostafa Hassan Mostafa [2,*] and Paul Terkumbur Adeke [2,3]

1 South African National Road Agency Limited (SANRAL), Eastern Region, Pietermaritzburg 3201, South Africa; naickers@nra.co.za

2 Sustainable Transportation Research Group (STRg), School of Engineering, University of KwaZulu-Natal, Durban 4041, South Africa; adekep@ukzn.ac.za, paul.adeke@uam.edu.ng or strg@ukzn.ac.za

3 Department of Civil Engineering, Joseph Sarwuan Tarka University, Makurdi 970101, Nigeria

* Correspondence: mostafam@ukzn.ac.za

Abstract: Transportation infrastructure underpins national mobility and economic growth, yet material sourcing for road construction imposes significant environmental and financial costs. As South Africa advances towards road construction, upcycling the reuse of reclaimed materials in higher-value applications offers opportunities to reduce waste and improve circular resource efficiency. This study assesses stakeholders' perception and adoption of upcycling in the Material Utilisation Plans (MUPs) for road construction. A mixed-methods approach combined nine semi-structured interviews and thirty-two survey responses from professionals involved in the National Route 3 upgrade project. Thematic analysis identified key qualitative themes, while quantituative data from a five-point Likert scale were examined through descriptive statistics, reliability, and correlation analysis. Respondents supported existing downcycling practices (mean = 3.682, SD = 1.088) and expressed readiness to adopt upcycling for pavement surfacing, base, subbase, and subgrade (mean > 3.00, SD < 1.30). Major barriers included client specifications, limited awareness and material cost constraints. Reliability analysis (Cronbach's $\alpha = 0.64–0.88$) confirmed internal consistency across qualitative themes. Also, there was a positive correlation between reclaimed materials and cost, design specifications, and optimised cost ($r > 0.30$, $p < 0.05$), while downcycling correlated negatively with costs ($r = -0.400$, $p < 0.05$). This study provides new empirical evidence on the systemic barriers hindering upcycling adoption in South African road projects and offers a validated mixed-method framework linking perceptual, technical, and economic dimensions of material reuse. It recommends integrating upcycling criteria into design, testing, and procurement processes, shifting from compliance-based recycling to performance-based circular material management in national road infrastructure.

Keywords: upcycling; materials utilisation plan; sustainable construction; road pavement; circular economy

1. Introduction

1.1. Background

Globally, the road construction sector is adopting circular economy strategies to enhance material efficiency, reduce costs, and mitigate environmental degradation [1–3]. Rapid urbanisation has intensified the demand for aggregates and bitumen while generating substantial waste [4]. Upcycling is the transformation of reclaimed materials into products of higher functional value, which offers a sustainable pathway for road construction, improving durability and cost-effectiveness compared with conventional recycling that typically downgrades material properties [2,5,6].

In South Africa, the uptake of upcycling remains limited due to inconsistent project planning, inadequate specification frameworks, and procurement systems that prioritise lowest-cost tendering over innovation [7–9]. The absence of explicit upcycling requirements in contract documents discourages contractors from innovation for fear of cost and risk penalties [7]. As a result, economic and environmental benefits are lost, despite evidence that well-processed reclaimed base and subbase materials can adequately substitute natural aggregates while maintaining efficient performance properties [4,10,11]. Embedding upcycling into project design could therefore lower emissions, conserve resources, and support long-term sustainability goals [12,13].

Another challenge is the limited awareness and technical expertise among designers, contractors, and decision-makers [9]. This knowledge gap extends to tertiary education, where upcycling concepts are rarely incorporated into pavement-engineering curricula. Without targeted capacity building, emerging professionals struggle to translate circular economy principles into practical design and construction choices [6,14].

Empirical evidence indicates that although awareness of recycling technologies is improving rapidly in recent times, the integration of upcycling within Materials Utilisation Plans (MUPs) is hindered by additional challenges, such as the lack of coordinated planning and risk-averse tendering processes [9,11,15–17]. Furthermore, international studies have highlighted the potential of recycled materials to achieve cost savings, lower greenhouse-gas emissions, and maintain structural integrity [4,11,13,18], underscoring the importance of systemic change within the infrastructure development system. No integrated study has recently linked design specifications, client requirements, and cost considerations to the practical implementation of upcycling within MUPs. This has left a major knowledge and practice gap between the technical feasibility demonstrated in the limited adoption observed in real-world projects.

Addressing this gap is therefore crucial to understanding how institutional culture, design frameworks, and procurement mechanisms influence the feasibility of upcycling within South African road construction projects.

Accordingly, this study investigates the institutional, technical, and economic factors influencing stakeholders' perceptions of upcycling within MUPs for road construction. By combining qualitative insights and quantitative validation, it develops an evidence-based framework for integrating upcycling principles into sustainable material management practices in South Africa's road sector.

The aim of this research is to examine stakeholders' perceptions and the factors influencing the adoption of upcycling practices in MUP design, as well as their implications for sustainable road infrastructure development. The research questions guiding this study are as follows:

1. What is the perception of South African experts in the road construction industry regarding upcycling in MUPs?
2. What are the factors influencing the specifications of upcycling in the design of MUPs in South Africa?
3. To what extent do the identified factors influence the specifications of upcycling in the design of MUPs in South Africa?

1.2. Theoretical Framework

Material optimisation in road construction requires the strategic integration of resource efficiency, cost management, and sustainability considerations across project life cycles [15,16]. The reuse of reclaimed materials from recycling—downcycling or upcycling has been demonstrated to reduce demand for natural resources, lower energy consumption, and decrease greenhouse-gas emissions [11,17]. The circular economy paradigm underpins this process by promoting a continuous flow of materials within construction systems rather than a linear extract–use–dispose model [1,2]. However, the growing demand for infrastructure continues to place significant strain on natural resources, highlighting both the opportunities and challenges of implementing sustainable material management practices [13].

Within this framework, MUPs serve as central instruments for coordinating how materials are sourced, processed, and reused within and between projects. An effective MUP provides an operational link between policy intent and project-level implementation, ensuring that reclaimed materials are systematically evaluated for potential reuse through upcycling in new pavement layers [9,10]. The interrelationship among these elements, including client requirements, design decisions, and economic viability, is represented in Figure 1, which illustrates the conceptual structure of a typical MUP.

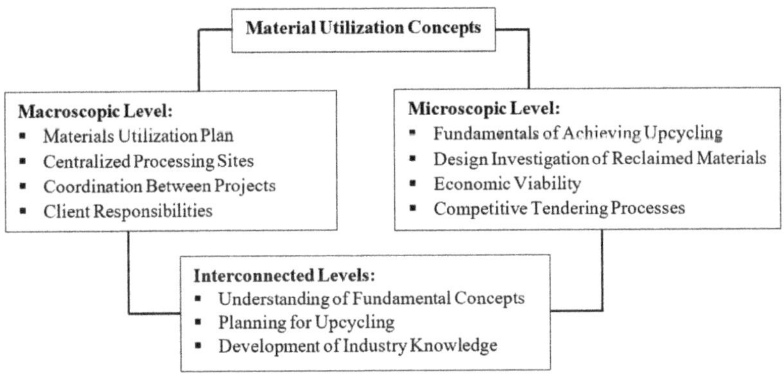

Figure 1. Conceptual schematic of a Material Utilisation Plan (MUP), showing the interaction between macroscopic, microscopic, and interconnected levels in upcycling and material management.

At a macroscopic level, MUPs support coordinated planning across multiple projects and regions, reducing duplication of quarrying and material transportation [9,19]. At a microscopic level, they address project-specific challenges such as local material variability, cost–benefit trade-offs, and testing constraints. In practice, this framework promotes collaboration between clients, designers, and contractors to optimise material flow, maintain uniform quality standards, and minimise environmental impacts [9,20]. The generic framework for recycling of road materials for an effective and sustainable material utilisation plan for road construction is presented in Figure 2.

Figure 2 depicts how the performance of reclaimed materials could be optimised through systematic cost and quality management strategies. It demonstrated that when the cost of natural materials remains higher than the properly processed reclaimed materials and when environmental externalities are accounted for, recycling–upcycling yields measurable economic and ecological advantages [4,21–23].

Technological advances such as geopolymerisation, cement stabilisation, and nanotechnology have further improved the structural and environmental performance of recycled and upcycled road materials [2,22–24]. However, the integration of these techniques into routine design and construction is constrained by the absence of standardised specifications, limited testing capacity, and misalignment between academic curricula and industry practice [9,21,25]. These institutional barriers underscore the need for coordinated reforms, particularly clearer design specifications for reclaimable materials, incentive frameworks for contractors, and updated engineering education programmes to mainstream upcycling within South Africa's road construction sector. In addition, the adoption of upcycling is influenced by the internal recycling and quality-control regulations of production companies, which remain largely unstandardised across the South African road materials industry. Most quarries and aggregate suppliers operate under environmental authorisations rather than performance-based recycling frameworks, thereby creating inconsistencies between private-sector production controls and public-sector design specifications. This regulatory misalignment limits the consistent incorporation of reclaimed materials into project-level MUPs and constrains coordination between producers and designers [3,6].

		To New Pavement				
		Surface Layer	Base Course	Subbase Layer	Selected Layer	Fill
From Existing Pavement	Surface Layer	Reuse	Downcycle	Downcycle	Downcycle	Downcycle
	Base Course	Upcycle	Reuse	Downcycle	Downcycle	Downcycle
	Subbase Layer	Upcycle	Upcycle	Reuse	Downcycle	Downcycle
	Selected Layer	Upcycle	Upcycle	Upcycle	Reuse	Downcycle

Key	Increased Value	Same Value	Decreased Value	Minimal Value

Figure 2. Matrix illustrating material transitions between pavement layers, indicating reuse, recycling, and downcycling processes. Colours reflect relative value change, from increased (upcycling) to minimal (downcycling) use.

Furthermore, project coordination across multiple stakeholders presents a persistent implementation challenge. Effective material management requires synchronised decision-making and transparent communication between clients, consultants, and contractors. However, fragmented institutional structures often result in inefficiencies and lost opportunities for reuse [4]. As Lavikka et al. [26] noted, digital platforms such as Building Information Modelling (BIM) can significantly enhance inter-project coordination, traceability, and performance monitoring. Despite these benefits, adoption remains limited due to cost, complexity, and organisational inertia, highlighting the need for systemic transformation in materials management practices [9].

Overall, this theoretical framework positions upcycling as both a technical innovation and an institutional reform process. It integrates circular economy theory, project management principles, and engineering design to explain how economic viability, client responsibility, and coordination interact to influence upcycling adoption in MUPs. In contexts such as South Africa, where material scarcity and infrastructure backlogs coexist, the framework provides a foundation for understanding and improving the technical feasibility and sustainability of road construction materials.

2. Materials and Methods

2.1. The Case Study

This study adopted a mixed-methods approach to investigate the institutional, economic, and technical factors influencing the adoption of upcycling within MUPs for a typical South African road project.

The process alleviates limitations of a single method [27]; however, it provides sufficient methods to investigate the intricacies and trends that respond to the research questions [28]. The approach combined semi-structured interviews with a structured questionnaire survey to triangulate perceptions from multiple professional categories or stakeholders involved in the National Route 3 (N3) upgrade project.

The N3 is integral to the economic growth of Southern Africa. A growing Southern African population has placed this strategic corridor under severe pressure, requiring substantial upgrading of portions of the corridor. This corridor, also referred to as the Durban-Free State–South African Government's key strategic integrated projects (SIP2) that forms part of the National Development Plan. The need for the project arose from increased traffic volumes resulting in declining levels of service on the N3 freeway. The freeway between Pietermaritzburg and Durban currently exceeds the South African National Road Agency Limited (SANRAL) permissible traffic density limit for the peak hour and the 30th highest hourly volume, respectively. Due to the substantial costs to upgrade the 80 km freeway, the route has been strategically split into thirteen reasonably sized work packages that form the collective N3 upgrade project. This corridor context provided an ideal testbed for examining upcycling adoption, as it represents the complex institutional, technical, and logistical structures typical of South Africa's major road projects. While the study draws data from one corridor, its strategic scope, diversity of stakeholders, and multi-package structure enhance the transferability of findings to other large-scale infrastructure programmes in similar contexts.

2.2. Research Design

This study adopted a convergent mixed-methods design, combining qualitative thematic analysis with quantitative descriptive and correlational analysis [29]. The design was chosen to integrate in-depth qualitative understanding with quantitative validation of the factors influencing the adoption of upcycling within MUPs. Qualitative methods were used to explore why certain barriers and enablers existed, while quantitative techniques tested how strongly these factors were associated with practice. Semi-structured interviews were conducted with nine key professionals engaged in project management, materials testing, and design engineering on the National Route 3 upgrade project. These interviews provided insight into institutional culture, decision-making processes, and perceived risks surrounding upcycling.

A structured questionnaire survey was then administered to thirty-two respondents drawn from the same professional ecosystem to measure the prevalence and strength of their perceptions across disciplines. The design ensured qualitative alignment between the identified themes and statistical constructs validated through Cronbach's Alpha and Pearson correlation coefficient values. The semi-structured interviews and questionnaires were used to collect both qualitative and quantitative data, respectively [30,31]. This flexible format allowed for modifications of questions based on the participants' responses, ensuring alignment with study objectives [32]. The integration of interview and survey data generated a coherent mixed-methods dataset that contributed to the validity and dependability of the study findings [32,33]. Within the context of the N3 upgrade project, which includes a corridor of multiple projects, this design facilitated the examination of both project-specific practices and broader institutional tendencies typical of South African road projects.

Although the empirical data were drawn from a single programme, the corridor's national strategic importance, multi-package structure, and diverse stakeholder composition strengthen the transferability rather than statistical generalisability of the study's conclusions to other large-scale infrastructure contexts.

2.3. Sampling and Representativeness

The study employed a purposive sampling approach within the defined project context rather than random selection across the national industry. This is appropriate when the goal is to obtain in-depth insights from participants with specific knowledge or decision-making experience relative to the research focus [31,34,35]. The selected participants, therefore, comprised stakeholders directly involved in project management, materials testing, and pavement design on the National Route 3 upgrade. While this approach limits the statistical generalisability of findings, it enhances their analytical depth and contextual validity by situating the results within a realistic and practice-based environment [35,36]. Consequently, the study positions its outcomes as analytically transferable rather than universally generalisable; that is, the insights and relationships identified are relevant to other infrastructure projects with comparable institutional and technical conditions [37].

To enhance balance and reduce bias within the purposively identified participant groups, a probability-informed approach consistent with standard methodological guidance [38] was applied during the survey phase. This ensured that all individuals within each identified category (e.g., project managers, materials engineers, and laboratory testers) had equal opportunity to participate, thereby improving representativeness while maintaining alignment with the study's purposive framework [39–41]. Being a mixed-methods research design, the qualitative aspect considered nine participants, while the quantitative approach used thirty-two participants from the target population. Although the final sample size appeared relatively small, it was acceptable for the study's scope and design [42,43]. In qualitative research, smaller, targeted samples were preferred to allow for in-depth exploration and thematic saturation [29], while on the other hand, the thirty-two respondents used for the quantitative method represented a substantial proportion of the targeted population, since the study focused on a specific and specialised workforce that represents the population under investigation.

2.4. Data Collection

Data collection procedures were designed to align with the mixed-methods framework, ensuring complementarity between the qualitative and quantitative strands.

The qualitative data collection process involved the use of voice recording platforms for recording and transcribing the responses based on stakeholders' perceptions on the adoption of upcycling for road construction in South Africa. Permissions were requested from the respondents before the start of the interview, and the activation of the recording and transcribing systems [42]. This approach minimised the duration of the interview [32]. It involved the development of the interview schedule, which allowed respondents the opportunity to provide more information related to the subject [33,34], carrying out pilot semi-structured interviews, performing the face-to-face interviews, and transcribing records for data analysis. The online and face-to-face interviews were scheduled with participants to create the opportunity to obtain relevant information [43]. Transcribed responses were reviewed and coded, from which key themes and subthemes were identified to capture participants' perspectives on the study objectives [38]. The results of the qualitative data collection approach were based on the direct experiences, opinions, and behaviour of human beings as meaning-making agents in their everyday lives. However, the respondents needed to understand the aim of the study as well as the research questions before granting the interviews [44].

The quantitative approach involved administering closed-ended research questionnaires to respondents to enable them to express their perceptions on the concept of upcycling in road construction in South Africa. The validation of the research instrument involved piloting it through an interview schedule with experts in the industry, aimed at reviewing it to confirm that all key subjects and elements relevant to answering the research questions were included [34]. The process gave insight into whether the interview schedule was clear, unambiguous, understandable, and concise enough for the study. This approach solicited quantifiable data from respondents to be analysed using statistical methods for objective and unequivocal decisions [31]. Efforts to minimise bias were ensured by inviting participants from all relevant professional categories, maintaining balance across roles and disciplines within the project. Major sections of the questionnaire were as follows: Part A for social demographic-related questions (age, gender, education, experience, and position occupied in the current industry), and the understanding of the MUPs and the various options to upcycle materials. Part B focused on the Laboratory testers and sought to determine the factors affecting the testing of materials. Part C applied to pavement design engineers and sought to understand the challenges with designing for upcycling, and Part D referred to all participants and sought to understand the factors affecting the MUPs. The questionnaire was structured using a five-point Likert scale, with response options defined as follows: 1 = Strongly Disagree, 2 = Disagree, 3 = Neutral, 4 = Agree, and 5 = Strongly Agree, ensuring comprehensive representation of respondents' levels of agreement [42].

In all approaches, ethical approval for the study was obtained from the University of KwaZulu-Natal, and informed consent was provided for all participants. Both instruments—the interview guide and questionnaire—were structured around the same conceptual framework, facilitating integration of qualitative, and quantitative findings during analysis.

2.5. Data Analysis

According to Braun and Clarke [45], thematic Analysis is a systematic technique for methodically identifying, categorising, and presenting insights into patterns of meaning, also referred to as themes, that are present in a dataset. The process involves six chronological steps:

Step 1: Data familiarisation, which entails transcription of respondents' responses by listening to the recording and writing down the key understandings.
Step 2: Generating the initial codes, which involves reviewing the transcript to identify initial codes based on comments that were relevant in the context of the research.
Step 3: Searching for themes—this involves the interpretation and categorisation of responses into themes and subthemes.
Step 4: Reviewing identified themes—completion of the initial search for themes, then scrutinising for coherence and appropriateness.
Step 5: Defining and naming themes—this involved the identification of themes, and iteratively refining and defining to correctly describe the essence of the themes for the study.
Step 6: Producing the report—by converting the themes into responses for the research questions. The final report was not limited to the description of the themes, but also included the analysis, which was supported by the literature and responded to the research questions [45].

Quantitative data were analysed using descriptive statistical parameters to explain frequency distributions and identify patterns within the dataset [46,47]. The essential parameters used were the sum, mean, and standard deviation, estimated as the arithmetic average or the representative value, and a measure of how far the data is spread around the mean or dispersion. The reliability test was employed to check the effectiveness of the quantitative research instrument. It attempted to test the suitability of the instrument for measuring variability and coherence among the variables, focusing on consistency. This confirms the internal consistency across all factors (Cronbach's Alpha Value $\alpha > 0.60$), and the closeness of individual questions related to each other [47]. Also, the Pearson correlation test was used to examine the relationship between factors influencing the adoption of upcycling among stakeholders at 5% significance level [48,49].

Results from both analytical strands were compared and merged during interpretation, allowing qualitative themes to clarify the underlying causes of quantitative patterns. This methodological triangulation strengthened the dependability, validity, and interpretive depth of the study's findings.

3. Results

3.1. Respondent Profile

The professional composition and experience distribution of respondents are summarised in Figure 3.

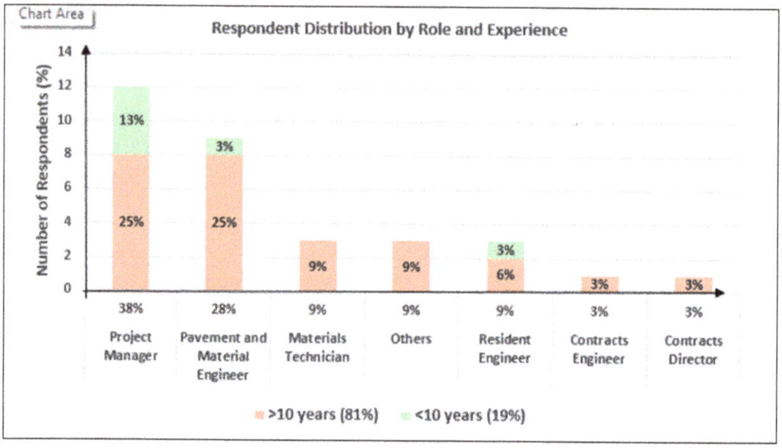

Figure 3. Respondent distribution by discipline and experience level, showing 81% had over 10 years of professional experience.

As illustrated in Figure 3, the sample comprised a diverse group of professionals involved in road construction project delivery, including project managers (38%), pavement and materials engineers (28%), and materials technicians (9%). Smaller proportions of respondents were resident engineers (9%), contract engineers (3%), and contract directors (3%), with a further 9% classified as "other" professional categories.

The experience profile of participants showed that 81% had more than ten years of industry experience, while 19% had less than ten years. This distribution indicated that the dataset predominantly reflects the perspectives of seasoned professionals with substantial exposure to materials management, design, and construction supervision. At the same time, the inclusion of early-career practitioners ensured representation of emerging viewpoints, providing a balanced and credible basis for interpreting stakeholder perceptions.

Overall, the composition of respondents mirrored the multidisciplinary structure of South Africa's road construction sector, thereby enhancing the contextual reliability and practical relevance of the findings [50,51].

3.2. Factors Affecting the Adoption of Upcycling in MUPs

3.2.1. Existing Culture of Road MUPs

According to the respondents, a typical MUP for a road construction project in South Africa usually assumes the trends shown in Table 1.

Table 1. Existing material recycling for road project MUPs.

Items	Questions	1	2	3	4	5	Total	Mean	SD
P1	The utilisation of reclaimed materials within a project is restricted to the project limits	9 28%	8 25%	2 6%	8 25%	5 16%	32 100%	2.75	1.50
P2	The utilisation of materials between adjacent projects is limited to surplus reclaimed material	4 13%	7 25%	2 19%	10 31%	9 22%	32 100%	3.41	1.43
P3	The movement of reclaimed materials between projects imposes a contractual risk	1 3%	8 25%	6 19%	10 31%	7 22%	32 100%	3.44	1.19
P4	It is economically viable to utilise a quarry owned by the client near the project	0 0%	0 0%	5 16%	11 34%	16 50%	32 100%	4.34	0.75
P5	It is economically viable to utilise a stockpile site owned by the client near the project	0 0%	0 0%	1 3%	15 47%	16 50%	32 100%	4.47	0.57
	Cluster Values							3.682	1.088

The results showed that reuse of reclaimed materials was generally confined within the individual project boundaries (mean = 2.75, SD = 1.50). However, the respondents strongly agreed that the use of client-owned quarries and stockpile sites located near project areas enhanced the overall cost efficiency (means > 4.3). In contrast, there was only moderate agreement regarding the reuse of surplus materials between adjacent projects and the perceived contractual risks associated with moving materials across sites (mean < 3.5).

The average of means in a given cluster or cluster mean of 3.682 (SD = 1.088) indicated the combined responses from subthemes reflecting responses to the major theme on material reuse behaviour and project-level decision factors, or the existing culture on material recycling for road project MUP. This indicates that, overall, logistical and contractual factors have a greater influence on reuse practices than the sustainability objectives. These findings are consistent with previous studies by Hoy et al. [4] and Wu et al. [52], who observed that operational constraints often take precedence over environmental considerations in material recovery and reuse.

3.2.2. Recycling for the Road Surface Layer

Results for the suitability of recycling existing road materials for the construction of new pavement layers, like the surface layer, are presented in Table 2.

Table 2. Recycling for road surface layer.

Items	Questions	1	2	3	4	5	Total	Mean	SD
P10	The reclaimed surface layer can be recycled and utilised in the surface layer of the new pavement	0 0%	7 22%	4 13%	8 25%	13 41%	32 100%	3.84	1.19
P11	The reclaimed surface layer can be downcycled and utilised in the new pavement	0 0%	2 6%	7 22%	11 34%	12 38%	32 100%	4.03	0.93
P12	The reclaimed surface layer can be downcycled and utilised as fill for the project	7 22%	6 19%	6 19%	7 22%	6 19%	32 100%	2.97	1.45
	Cluster Values							3.623	1.19

Respondents generally supported the recycling and downcycling of reclaimed surface layers. They agreed that such materials could be recycled and reused in the surface layer of new pavements (mean = 3.84, SD = 1.19) and downcycled for use in underlying pavement layers (mean = 4.03, SD = 0.93). However, there was a weaker agreement for using the surfacing layer as fill (mean = 2.97, SD = 1.45), suggesting that respondents perceive greater value in reusing the surface materials for higher-order pavement applications rather than for low-value fill.

The overall cluster mean of 3.623 (SD = 1.19) reinforced this trend, indicating a broad acceptance of recycling and downcycling practices for reclaimed surface materials in road construction. This agreement reflects the growing acceptance of reuse in upper pavement layers [10,15].

3.2.3. Recycling for Base Layer

Results for the perception of respondents on recycling for the production of the base layer for new road pavement construction are presented in Table 3.

Table 3. Considerations for producing base layer.

Items	Questions	1	2	3	4	5	Total	Mean	SD
P13	The reclaimed base layer can be upcycled and utilised in the surface layer of the new pavement	10 31%	8 25%	8 25%	4 13%	2 6%	32 100%	2.38	1.24
P14	The reclaimed base layer can be recycled and utilised in the base layer of the new pavement	2 6%	3 9%	12 38%	7 22%	8 25%	32 100%	3.50	1.16
P15	The reclaimed base layer can be downcycled and utilised in the new Pavement	0 0%	1 3%	5 16%	13 41%	13 41%	32 100%	4.19	0.82
P16	The reclaimed base layer can be downcycled and utilised as fill for the project	5 16%	4 13%	9 28%	6 19%	8 25%	32 100%	3.25	1.39
	Cluster Values							3.330	1.153

Respondents disagreed with the practice of upcycling the existing base layer for use in surface courses (mean = 2.38, SD = 1.24), but supported recycling within the base layer itself (mean = 3.50, SD = 1.16) and downcycling into lower pavement layers (mean = 4.19, SD = 0.82). Qualitative responses indicated that engineers viewed base materials as technically stable but not sufficiently tested for higher-order functional reuse.

The cluster mean of 3.330 (SD = 1.15), which was slightly above the neutral limit (3.0), reflected a general cautious approach, showing concerns about performance. These results echoed stakeholder scepticism reported in prior studies [10], where uncertainty around testing and specifications limited the willingness to upcycle base materials into road surface applications.

3.2.4. Recycling for the Subbase Layer

Results of investigations into the acceptance of upcycling for the recycled subbase layer for road construction in South Africa are presented in Table 4.

Upcycling of the subbase layer was met with mixed opinions (mean = 2.72, SD = 1.33). Respondents showed moderate agreement with recycling within the subbase (mean = 3.47, SD = 1.27) and downcycling into fill or lower layers (means = 3.81 and 3.53; SD = 1.15 and 1.29). Stakeholders generally supported materials reuse, but not necessarily upcycling, indicating a continued preference for recycling and reuse rather than value-added upcycling. The cluster mean of 3.383 (SD = 1.26) underscored this conservative approach and highlights the need for clearer performance criteria and improved testing protocols to build confidence in subbase upcycling for sustainable pavement design [10].

Table 4. Considerations for producing subbase layer.

Items	Questions	1	2	3	4	5	Total	Mean	SD
P17	The reclaimed subbase layer can be upcycled and utilised in the new pavement	9 28%	2 6%	14 44%	3 9%	4 13%	32 100%	2.72	1.33
P18	The reclaimed subbase layer can be recycled and utilized in the subbase layer of the new pavement	3 9%	3 9%	11 32%	6 19%	9 28%	32 100%	3.47	1.27
P19	The reclaimed subbase layer can be downcycled and utilised in the new pavement	1 3%	4 13%	6 19%	10 31%	11 34%	32 100%	3.81	1.15
P20	The reclaimed subbase layer can be downcycled and utilised as fill for the project	2 6%	6 19%	7 22%	7 22%	10 31%	32 100%	3.53	1.29
	Cluster Values							3.383	1.26

3.2.5. Considerations for Subgrade Layers

The respondents' perception of recycling some subgrade layers is presented in Table 5.

Table 5. Subgrade layer.

Items	Questions	1	2	3	4	5	Total	Mean	SD
P21	The reclaimed Subgrade layer can be upcycled and utilised in the new pavement	10 31%	6 19%	11 34%	1 3%	4 13%	32 100%	2.47	1.32
P22	The reclaimed Subgrade layer can be recycled and utilised in the Subgrade layers of the new pavement	1 3%	2 6%	8 25%	11 34%	10 31%	32 100%	3.84	1.05
P23	The reclaimed Subgrade layer can be downcycled and utilised as fill for the project	3 9%	1 3%	4 13%	12 38%	12 38%	32 100%	3.91	1.23
	Cluster Values							3.407	1.20

Respondents did not support upcycling subgrade materials for the higher layer (mean = 2.47, SD = 1.32), but agreed with recycling and downcycling for fill and subgrade reuse (means = 3.84 and 3.91). The findings suggest a risk-averse engineering culture, where stability and proven performance outweigh sustainability objectives. Reuse acceptance decreased progressively from surface to subgrade layers, reflecting confidence in recycling only when risk is minimal. The cluster mean of 3.407 (SD = 1.20) indicates limited acceptance of upcycling and continued caution among practitioners.

3.2.6. Client's Specifications

The client's specifications and design requirements influenced the perception of road experts on material utilisation through the integration of upcycling in the MUP for road construction in South Africa, as shown in Table 6.

Table 6. The specifications influencing materials utilisation.

Items	Questions	1	2	3	4	5	Total	Mean	SD
P24:	The clients' specifications are a barrier to the recycling of materials	5 16%	5 16%	5 16%	10 31%	7 22%	32 100%	3.28	1.40
P25:	It is the responsibility of the client to specify the minimum % of reclaimable material to be used in each layer of the new pavement	5 16%	5 16%	5 16%	10 31%	7 22%	32 100%	3.28	1.39
P26:	There is limited awareness of the need for upcycling of the existing pavement layers	0 0%	2 6%	1 3%	16 50%	13 41%	32 100%	4.25	0.80
P27:	The engineer is responsible for designing the upcycling of the existing pavement layers	0 0%	3 9%	6 19%	9 28%	14 44%	32 100%	4.06	1.01
P28:	The current testing regime during the design phase is insufficient to inform the upcycling of the existing pavement layers	2 6%	2 6%	9 28%	14 44%	5 16%	32 100%	3.56	1.05
	Cluster Values							3.686	1.13

Table 6 revealed that respondents viewed client specifications as a barrier to recycling (mean = 3.28, SD = 1.40) but strongly agreed that engineers should design for upcycling (mean = 4.06, SD = 1.01). There was strong agreement that limited awareness and inadequate design responsibility hindered the adoption of upcycling (mean > 4.00, SD < 1.2). Also, the current testing regimes were insufficient to guide design decisions; nevertheless, testing significantly influenced the adoption (mean = 3.56, SD = 1.05). Although engineers support the concept of reuse, tender and contract documents seldom specify measurable reuse targets. Therefore, the cluster values (mean = 3.686, SD = 1.13) justified the assertion that institutional awareness, specification clarity, and testing limitations constrain the adoption of upcycling. These findings highlight ongoing uncertainty regarding long-term performance and the absence of a standardised testing framework reported by Jamshidi and White [51]. This implied that stakeholders recognised that upcycling could help reduce depletion of natural aggregates, but practical adoption remained limited by institutional and technical constraints or design specifications, which conformed with findings in previous studies on sustainable pavement materials [2,3,15]. This gap mirrors governance challenges identified by Gobbo et al. [7], thereby suggesting the need for clearer policy direction.

3.2.7. Material Cost

The influence of material cost on the integration of upcycling in MUP was examined using the responses obtained from the survey. Results of the analysis are presented in Table 7.

Table 7. Cost of natural resources in a flexible pavement project.

Items	Questions	1	2	3	4	5	Total	Mean	SD
P29:	The process of recycling has cost implications, and it is easier to spoil materials	11 34%	7 22%	9 28%	3 9%	2 6%	32 100%	2.31	1.23
P30:	The process of upcycling has cost implications, and it is easier to downcycle materials	4 13%	3 9%	11 34%	7 22%	7 22%	32 100%	3.31	1.28
P31:	Contractors need to be incentivised to reclaim and upcycle the existing pavement layers	1 0%	3 6%	7 3%	12 50%	9 41%	32 100%	3.78	1.07
P32:	The introduction of a surcharge when spoiling material at a landfill site will increase the efficient utilisation of existing pavement layers in the new pavement	3 9%	1 3%	8 25%	10 31%	10 31%	32 100%	3.71	1.22
	Cluster Values							3.278	1.20

Material cost was recognised as a key determinant (Table 7). Respondents disagreed that spoiling materials was cost-effective (mean = 2.31, SD = 1.23) but supported financial incentives (mean = 3.78, SD = 1.07) and landfill surcharges (mean = 3.71, SD = 1.22) to encourage reuse. These findings emphasise that economic drivers strongly influenced material utilisation decisions. Qualitative findings indicated that introducing financial incentives such as rebates for recycled content could redirect contractor behaviour toward upcycling. The cluster mean of 3.279 (SD = 1.20) supports the conclusion that financial incentives could play a vital role in promoting sustainable material reuse.

3.3. Reliability Test

The reliability test was used to assess the strength of the decision variables employed in the study. According to Gravetter and Wallnau [50], an Alpha Coefficient Range of 0.6 to <0.70 strength of association could be regarded as moderate, 0.7 to <0.8 as good, and >0.8 as very good. Cronbach's Alpha findings of the study variables are presented in Table 8.

Table 8. Reliability test of decision variables.

Study Variables	Cronbach's Alpha	Counts
Factor 1—Reusing the existing reclaimed materials	0.88	7
Factor 2—Downcycling material to fill	0.86	4
Factor 3—Limited design for upcycling	0.77	5
Factor 4—Optimising costs	0.79	5
Factor 5—Client's responsibilities	0.81	5
Factor 6—Reluctance to upcycle	0.64	4
Factor 7—Economic viability of materials utilisation	0.66	3

Cronbach's Alpha values ranged from 0.64 to 0.88, indicating moderate to high internal consistency across the measured factors, confirming that the survey items were reliable for this study and relevant for subsequent correlation analysis.

3.4. Correlation Analysis

Pearson's Product-Moment correlation analysis was used to examine the relationship between the variables in terms of proportionality, such that when one variable changes in value, the other tends to change in a specific direction, either positive or negative. The magnitude of the change was measured using Pearson's (r) correlation, which ranged between -1 for a perfect negative correlation to +1 for a perfect positive correlation at 5% level of significance [45]. The correlation benchmarks were defined as $0.10 \leq r \leq 0.29$ for small, $0.30 \leq r \leq 0.49$ for medium, and $0.50 \leq r \leq 1.00$ for large effect. The correlation between influencing factors considered by this study is presented in Table 9.

Table 9. Correlation analysis of influencing factors.

	Factor 1:	Factor 2:	Factor 3:	Factor 4:	Factor 5:	Factor 6:	Factor 7:
Factor 1:	1						
Factor 2:	0.198	1					
Factor 3:	0.302	−0.221	1				
Factor 4:	0.330	−0.117	0.142	1			
Factor 5:	0.209	−0.146	−0.064	−0.007	1		
Factor 6:	0.030	−0.048	0.296	0.136	−0.042	1	
Factor 7:	0.353	−0.400	0.209	0.193	0.174	0.108	1

These results suggest that cost-sensitive projects with clear client guidance were more likely to incorporate reclaimed materials. Interview data supported this claim, noting that consistent specification and incentive frameworks could encourage upcycling practices. Similar mechanisms were advocated by Singh and Gupta [24] for promoting the adoption of the circular economy. A positive relationship between Factor 1 (Reusing existing reclaimed materials) and Factor 4 (Optimising costs) ($r = 0.330$) indicated that the respondents associated material reuse with cost-saving opportunities. Similarly, Factor 1 also correlated moderately with Factor 7 (Economic viability of material utilisation) ($r = 0.353$), suggesting that greater reuse aligns with perceptions of economic efficiency.

The weak positive correlation between Factor 3 (Limited design for upcycling) and Factor 6 (Reluctance to upcycle) ($r = 0.296$) implied that where design limitations were perceived, the hesitation to implement upcycling increased, reinforcing the importance of technical guidance and design standards.

Conversely, the negative correlation between Factor 2 (Downcycling material to fill) and both Factor 3 (Limited design for upcycling) ($r = -0.221$) and Factor 7 (Economic viability) ($r = -0.400$) suggested that a stronger preference for downcycling may undermine innovation and economic efficiency. In practice, this indicates that defaulting to low-value reuse (fill) reflects lost economic and environmental opportunities. These insights emphasise that improving design capability and financial incentives could strengthen the perceived viability of sustainable material utilisation.

In summary, these findings highlight that, while technical feasibility and moderate willingness exist, the broader perception of stakeholders on the adoption of upcycling is constrained by institutional inertia, clients' and designs' specification rigidity, and perceived contractual risks.

4. Conclusions

This mixed-methods case study provided empirical evidence that the adoption of upcycling in the South African road construction industry was constrained more by institutional and procedural barriers than by technical limitations. Competitive tendering, limited testing capacity, and insufficient design-stage coordination emerged as the most critical obstacles. Correlation analysis confirmed that material reuse and cost optimisation were positively related ($r = 0.353$), while downcycling showed a moderate negative correlation with cost ($r = -0.400$), indicating that cost perceptions significantly influence materials use choice. Design frameworks ($r = 0.302$) and institutional policies ($r = 0.330$) were also positively linked to upcycling adoption, showing that coordinated planning and clear governance improve decision outcomes.

A strong consensus emerged among stakeholders regarding the long-term cost and environmental benefits of upcycling when supported by clear client specifications and policy incentives. Embedding upcycling requirements into design guidelines could therefore advance material efficiency and carbon reduction in South Africa's road sector.

Although the study was limited in sample size, it provides a validated diagnostic framework that can inform future researchers and policymakers on circular material feasibility. By framing upcycling within a practical policy, procurement, and educational framework, this study supports South Africa's transition towards a sustainable, resource-efficient, and low-carbon road construction industry.

Author Contributions: Conceptualisation, M.M.H.M. and S.N.; Methodology, M.M.H.M., S.N. and P.T.A.; Validation, M.M.H.M., S.N. and P.T.A.; Formal Analysis, S.N. and P.T.A.; Investigation, S.N.; Resources, S.N.; Data Curation, S.N.; Writing—Original Draft Preparation, S.N.; Writing—Review and Editing, M.M.H.M., S.N. and P.T.A.; Visualisation, S.N. and P.T.A.; Supervision, M.M.H.M.; Project Administration, M.M.H.M.; Funding Acquisition, M.M.H.M. All authors have read and agreed to the published version of the manuscript.

Funding: This research was partially funded by the KwaZulu-Natal Department of Transportation (KZNDoT) Chair in Sustainable Transportation at the University of KwaZulu-Natal, Durban, South Africa.

Institutional Review Board Statement: The study was approved by the Humanities & Social Sciences Research Ethics Committee (UKZN Research Ethics), HSSREC/00003378, on 6 October 2022.

Informed Consent Statement: Informed consent was obtained from all subjects involved in the study.

Data Availability Statement: The original contributions presented in this study are included in the article. Further inquiries can be directed to the corresponding author.

Acknowledgments: The authors are indebted to the KZNDoT and SANRAL for their support. We also gratefully acknowledge the support and contributions of industry stakeholders who assisted in this study.

Conflicts of Interest: Author Salome Naicker was employed by the company South African National Road Agency Limited (SANRAL). The remaining authors declare that the research was conducted in the absence of any commercial or financial relationships that could be construed as a potential conflict of interest.

References

1. Gnatiuk, L.; Novik, H.; Melnyk, M. Recycling and upcycling in construction. *Theory Pract. Des.* **2022**, *1*, 130–139. [CrossRef]
2. Galila, A.; Kaseireldeil, S.; Abdalkhaliq, N.; Farouk, E.; Eichner, M.; Sarhan, Y. Enhancing sustainability and resource efficiency through upcycling: A comprehensive review and analytical–based framework for evaluating building upcycled products. *Innov. Infrastruct. Solut.* **2024**, *9*, 270. [CrossRef]
3. Liu, Z.; Kringos, N. Transition from linear to circular economy in pavement engineering: A systematic review. *J. Clean. Prod.* **2024**, *460*, 141957. [CrossRef]
4. Hoy, M.; Horpibulsuk, S.; Chinkulkijniwat, A.; Suddeepong, A.; Buritatum, A.; Yaowarat, T.; Choenklang, P.; Udomchai, A.; Kantatham, K. Innovations in recycled construction materials: Paving the way towards sustainable road infrastructure. *Front. Built Environ.* **2024**, *10*, 1449970. [CrossRef]

5. Sung, K. Understanding Upcycling and Circular Economy and Their Interrelationships Through Literature Review for Design Education. In Proceedings of the Design Society: Design Conference 2023, Dubrovnik, Croatia, 22–25 May 2023; pp. 1347–1356. [CrossRef]
6. Detterborn, T.; Korkiala-Tanttu, L. Environmentally sustainable use of recycled crushed concrete aggregate in earthworks. In Proceedings of the 19th International Conference on Soil Mechanics and Geotechnical Engineering, Seoul, Republic of Korea, 17–22 September 2017.
7. Gobbo, E.; Nia, E.M.; Straub, A.; Stephan, A. Exploring the effective reuse rate of materials and elements in the construction sector. *J. Build. Eng.* **2024**, *98*, 111344. [CrossRef]
8. Van Dam, T.J.; Harvey, J.; Muench, S.T.; Smith, K.D.; Snyder, M.B.; Al-Qadi, I.L.; Ozer, H.; Meijer, J.; Ram, P.; Roesler, J.R.; et al. *Towards Sustainable Pavement Systems: A Reference Document*; Federal Highway Administration: Washington, DC, USA, 2015.
9. Naicker, S. The Feasibility of a Network Materials Utilisation Plan, with an Emphasis on Upcycling of Materials. Master's Thesis, University of KwaZulu-Natal, Durban, South Africa, 2022.
10. Abdalla, A.; Alsalihi, M.; Khafagy, M.; Ibrahim, H. Upcycled waste for sustainable roads: Exploring the synergistic effects of off-spec fly ashes and rejuvenators on asphalt binder aging. *Constr. Build. Mater.* **2025**, *467*, 140377. [CrossRef]
11. Václav, T.; Dana, M. Use of Recycled Materials in Road Construction. *J. Bus. IT* **2021**, *11*, 54–60. [CrossRef]
12. Zhao, Y.; Goulias, D.; Peterson, D. Recycled Asphalt Pavement Materials in Transport Pavement Infrastructure: Sustainability Analysis & Metrics. *Sustainability* **2021**, *13*, 8071. [CrossRef]
13. Harvey, J.; Meijer, J.; Ozer, H.; Al-Qadi, I.L.; Saboori, A.; Kendall, A.L. *Pavement Life-Cycle Assessment Framework USA*; Federal Highway Administration: Washington, DC, USA, 2016.
14. Milad, A. Recycled and upcycled materials in contemporary architecture and civil engineering: Their applications, benefits, and challenges. *Clear. Waste Syst.* **2025**, *10*, 100203. [CrossRef]
15. Yaro, N.S.A.; Sutanto, M.H.; Baloo, L.; Habib, N.Z.; Usman, A.; Yousafzai, A.K.; Ahmad, A.; Birniwa, A.H.; Jagaba, A.H.; Noor, A.A. Comprehensive Overview of the Utilisation of Recycled Waste Materials and Technologies in Asphalt Pavements: Towards Environmental and Sustainable Low-Carbon Roads. *Processes* **2023**, *11*, 2095. [CrossRef]
16. Liu, H.J.; Love, P.E.; Sing, M.C.; Niu, B.; Zhao, J. Conceptual framework of lifecycle performance measurement: Ensuring the resilience of transport infrastructure assets. *J. Transp. Res. Part D-Transp.* **2019**, *77*, 615–626. [CrossRef]
17. Baliello, A.; Wang, D. Advances in Road Engineering: Innovation in Road Pavements and Materials. *Buildings* **2024**, *14*, 2250. [CrossRef]
18. Rasch, M.K. Applying Material Flow Analysis for Optimizing Construction Aggregates Management in the Road Sector. Master's Thesis, Norwegian University of Science and Technology, Trondheim, Norway, 2018.
19. Sung, K. A review on upcycling: Current body of literature, knowledge gaps and a way forward. In Proceedings of the ICECESS 2015, 17th International Conference on Environmental, Cultural, Economic and Social Sustainability, Venice, Italy, 13–14 April 2015.
20. Ross, D.; Townsend, M. The Road Maintenance Backlog in South Africa. In Proceedings of the 38th Annual Southern African Transport Conference, Pretoria, South Africa, 8–11 July 2019.
21. Aziz, R.F.; Abdel-Hakam, A.A. Exploring Delay Causes of Road Construction Projects in Egypt. *Alex. Eng. J.* **2016**, *55*, 1515–1539. [CrossRef]
22. Jordaan, G.; Kilian, A.M.P.D.L.; Murphy, M. The Development of Cost-Effective Pavement Design Approaches Using Minerology Tests with New Nano-Technology Modifications of Materials. In Proceedings of the 36th Southern African Transport Conference (SATC 2017), Pretoria, South Africa, 10–13 July 2017.
23. Jordaan, G.; Kilian, A. The cost-effective upgrading, preservation and rehabilitation of roads-optimising the use of available technologies. In Proceedings of the 35th Southern African Transport Conference, Pretoria, South Affrica, 4–7 July 2016.
24. Singh, A.; Gupta, A. Upcycling of plastic waste in bituminous mixes using dry process: Review of laboratory to field performance. *Constr. Build. Mater.* **2024**, *425*, 136005. [CrossRef]
25. Li, N.; Mo, L.; Unluer, C. Emerging CO_2 utilisation technologies for construction materials: A review. *J. CO2 Util.* **2022**, *65*, 102237. [CrossRef]
26. Lavikka, R.H.; Smeds, R.; Jaatinen, M. Coordinating collaboration in contractually different complex construction projects. *Supply Chain. Manag. Int. J.* **2015**, *20*, 205–221. [CrossRef]
27. Dawidi, S.; Shrestha, S.; Giri, R.A. Mixed-Methods: Research: A Discussion on its Types, Challenges, and Criticisms. *J. Pract. Stud. Educ.* **2021**, *2*, 25–36. [CrossRef]
28. Onwuegbuzie, A.J.; Collins, K.M.T.; Leech, N.L.; Jiao, Q.G. Mixed data collection and analysis for conducting research on giftedness and beyond. In *Methodologies for Conducting Research on Giftedness*; American Psychological Association: Washington, DC, USA, 2010.
29. Creswell, J.W.; Poth, C.N. *Qualitative Inquiry and Research Design: Choosing Among Five Approaches*; SAGE Publications, Inc.: Los Angeles, CA, USA, 2018.

30. Creswell, J.W.; Creswell, J.D. *Research Design: Qualitative, Quantitative, and Mixed Methods Approaches*; SAGE Publications, Inc.: Los Angeles, CA, USA, 2018.
31. Berndt, A.E. Sampling Methods. *J. Hum. Lact.* **2020**, *36*, 224–226. [CrossRef] [PubMed]
32. Saunders, M.; Lewis, P.; Thornhill, A. *Research Methods for Business Students*; Pearson: Harlow, UK, 2016.
33. Barrett, D.; Twycross, A. Data collection in qualitative research. *Evid. Based Nurs.* **2018**, *21*, 63–64. [CrossRef]
34. Zorabi, M. Mixed Method Research: Instruments, Validity, Reliability and Reporting Findings. *Theory Pract. Lang. Stud.* **2013**, *3*, 254–262. [CrossRef]
35. Patton, M.Q. *Qualitative Research & Evaluation Methods, 4th ed*; SAGE Publications: Thousand Oaks, CA, USA, 2015.
36. Onwuegbuzie, A.J.; Johnson, R.B. The status of mixed methods research: Past, present, and future. In *The SAGE Handbook of Mixed Methods Research in Social & Behavioral Sciences*, 3rd ed.; Edwards, T.C., Ed.; SAGE Publications: Thousand Oaks, CA, USA, 2021; pp. 63–98.
37. Creswell, J.W.; Plano Clark, V.L. *Designing and Conducting Mixed Methods Research*, 3rd ed.; SAGE Publications: Thousand Oaks, CA, USA, 2018.
38. Pandey, P.; Pandey, M.M. *Research Methodology: Tools and Techniques*; Bridge Centre: Bucharest, Romania, 2015.
39. Etikan, I.; Bala, K. Sampling and Sampling Methods. *Biostat. Int. J.* **2017**, *5*, 215–217. [CrossRef]
40. Etikan, I. Comparison of Convenience Sampling and Purposive Sampling. *Am. J. Theor. Appl. Stat.* **2016**, *5*, 1–4. [CrossRef]
41. Young, T.J. *Questionnaires and Surveys*; Wiley Blackwell: Oxford, UK, 2016.
42. Aspers, P.; Corte, U. What is Qualitative Research? *Qual. Sociol.* **2021**, *44*, 599–608. [CrossRef]
43. Saarijärvi, M.; Bratt, E.L. When face-to-face interviews are not possible: Tips and tricks for video, telephone, online chat, and email interviews in qualitative research. *Eur. J. Cardiovasc. Nurs.* **2021**, *20*, 392–396. [CrossRef]
44. De Villiers, C.; Farook, M.B.; Molinari, M. Qualitative Research Interviews Using Online Video Technology-Challenges and Opportunities. *Meditari Account. Res.* **2021**, *30*, 1764–1782. [CrossRef]
45. Braun, V.; Clarke, V. Using thematic analysis in psychology. *Qual. Res. Psychol.* **2006**, *3*, 77–101. [CrossRef]
46. Bryman, A.; Bell, E. *Business Research Methods*, 6th ed.; Oxford University Press: Oxford, UK, 2023; ISBN 978-0-19-886944-2.
47. Field, A. *Discovering Statistics Using IBM SPSS Statistics*, 6th ed.; Sage Publications: London, UK, 2021; ISBN 978-1-5297-7018-2.
48. Vogt, W.P. Selecting the right analyses for your data. *Br. J. Educ. Technol.* **2015**, *46*, E4–E5. [CrossRef]
49. Pallant, J. *SPSS Survival Manual*; Routledge: London, UK, 2020.
50. Gravetter, F.J.; Wallnau, L.B. *Statistics for the Behavioral Sciences*, 11th ed.; Cengage Learning: Boston, MA, USA, 2021; ISBN 978-0-357-13441-8.
51. Jamshidi, A.; White, G. Use of recycled materials in pavement construction for environmental sustainability. In Proceedings of the Eighteenth Annual International Conference on Pavement Engineering, Asphalt Technology and Infrastructure, Liverpool, UK, 27–28 February 2019.
52. Wu, J.; Ye, X.; Cui, H. Recycled Materials in Construction: Trends, Status, and Future of Research. *Sustainability* **2025**, *17*, 2636. [CrossRef]

MDPI AG
Grosspeteranlage 5
4052 Basel
Switzerland
Tel.: +41 61 683 77 34

Buildings Editorial Office
E-mail: buildings@mdpi.com
www.mdpi.com/journal/buildings

www.ingramcontent.com/pod-product-compliance
Lightning Source LLC
Chambersburg PA
CBHW061622210926
51850CB00009B/2407

* 9 7 8 3 7 2 5 8 6 6 9 8 4 *